TRANSACTIONS

OF THE

AMERICAN PHILOSOPHICAL SOCIETY

HELD AT PHILADELPHIA
FOR PROMOTING USEFUL KNOWLEDGE

NEW SERIES—VOLUME 50, PART 6
1960

THE PALEOCENE PANTODONTA

ELWYN L. SIMONS

Department of Geology, Princeton University

THE AMERICAN PHILOSOPHICAL SOCIETY
INDEPENDENCE SQUARE
PHILADELPHIA 6

JULY, 1960

Library of Congress Catalog
Card No. 60–13623

THE PALEOCENE PANTODONTA

Elwyn L. Simons *

CONTENTS

PAGE

I. Introduction................................. 3
    Acknowledgments............................. 4
    Abbreviations of institutions..................... 4

II. History of study............................. 4

III. Systematic revision........................... 5
    Introduction................................ 5
    Pantodonta................................. 8
        Coryphodontoidea......................... 8
        Coryphodontidae......................... 8
            Coryphodon........................... 11
        Pantolambdoidea.......................... 15
        Pantolambdidae.......................... 15
            Pantolambda......................... 16
            Caenolambda........................ 21
        Barylambdidae........................... 25
            Barylambda.......................... 27
            Haplolambda......................... 27
            Leptolambda......................... 28
            Ignatiolambda....................... 29
        Titanoideidae........................... 31
            Titanoides........................... 33

IV. The pantodont skeleton....................... 40
    Introduction................................ 40
    Skull...................................... 40
    Permanent dentition......................... 42

Milk dentition............................... 43
Mandible................................... 45
Hyoid arch................................. 46
Vertebrae.................................. 46
Ribs....................................... 50
Clavicle.................................... 50
Scapula.................................... 50
Humerus................................... 51
Radius..................................... 51
Ulna....................................... 53
Manus..................................... 54
Pelvis..................................... 57
Femur..................................... 58
Patella.................................... 59
Tibia...................................... 59
Fibula..................................... 60
Pes....................................... 60

V. Geologic and geographic occurrence.............. 63

VI. Paleoecology................................. 68

VII. Relationships to other orders................... 70
    Appendix.................................. 73
        A. Logarithm of ratio diagrams............ 73
        B. Measurements of pantodont dentitions....... 73
    References.................................. 80

I. INTRODUCTION

Although the first pantodont to be discovered, apparently a specimen of *Coryphodon*, was found almost a hundred and fifty years ago in France, the rate of discovery of new genera and species of Pantodonta was not high before 1900. Within the last thirty years more pantodont genera have been proposed than in the whole of the previous century. Considering the widespread efforts of early collectors it is unusual that so many of these largest of Paleocene mammals should have escaped detection so long.

During the 1930's an extensive series of Paleocene pantodont skeletal material was collected in Mesa County, Colorado, by Professor Bryan Patterson and field parties of the Chicago Natural History Museum. This material, which included some of the most complete Paleocene mammalian skeletons ever discovered, made necessary a reconsideration of the taxonomy and anatomy of the Pantodonta. This revision reached its most recent expression in Patterson (1939a).

The present study was initiated by the collection of

many new pantodont specimens over the past twenty-nine years by Princeton University field parties under the direction of Dr. G. L. Jepsen in the Bighorn Basin of Wyoming, in Fallon and Carter Counties, Montana, and at the type locality of *Titanoides primaevus* near Fort Buford, North Dakota. Recovery of pantodonts during the last ten years in Big Horn County, Wyoming, to the southeast of the Polecat Bench has been particularly extensive and has yielded more or less fragmentary remains of over fifty individuals. These new materials and others now available make possible and desirable a fuller analysis of the Pantodonta than has been previously attempted.

The procedure of this paper will be to present first a taxonomic review of the Pantodonta with a description of new genera and species, and then a discussion of the skeletal anatomy of the various pantodonts, their stratigraphic occurrence, their ecology (as reflected by their sedimentary environments of fossilization and by their anatomy), and finally a consideration of the relationships of the Pantodonta to other orders of mammals.

* Present address of author: Zoological Laboratory, University of Pennsylvania, Philadelphia.

3

ACKNOWLEDGMENTS

Recognition is given here to the generosity of Dr. G. L. Jepsen in making available for study the extensive collection of undescribed Paleocene pantodonts at Princeton University assembled through his efforts, and for his constructive advice and helpful guidance in the course of preparing this work. Many of the problems relating to taxonomy and stratigraphic occurrence have been profitably discussed with Professor Bryan Patterson of Harvard University, and particular thanks are due to him for making available for study the excellent series of Paleocene pantodonts at the Chicago Natural History Museum collected under his direction. Dr. G. G. Simpson of the American Museum of Natural History has also been most generous in releasing, for description here, the type of *Ignatiolambda barnesi*, and in turning over for use in this work a manuscript pertaining to *Ignatiolambda*, as well as an illustration of the dentition of this form.

Mr. Chester Tarka, of the American Museum of Natural History staff, has given advice on the preparation of some of the illustrations included here. He is also to be thanked for preparing figure 12.

Financial assistance provided from the William Berryman Scott Research Fund of Princeton University made possible the examination of virtually all Paleocene pantodont specimens and helped to defray the cost of illustrations.

Appreciation for cooperation and efforts in lending pantodont specimens from other institutions is expressed to the following:

Dr. W. D. Turnbull, Chicago Natural History Museum, for the types of *Haplolambda quinni* Patterson and *Titanoides primaevus* [= *Sparactolambda looki*, Patterson], and specimens of *Leptolambda schmidti* Patterson and Simons and *Barylambda faberi* Patterson.

Dr. C. L. Gazin, United States National Museum, for specimens of *Pantolambda bathmodon* and *Pantolambda cavirictus* and the types of *Caenolambda pattersoni* Gazin, *Titanoides primaevus* Gidley, *Pantolambda intermedius* Simpson.

Dr. George Gaylord Simpson, American Museum of Natural History, for many specimens, both from the general collections and in the Cope Collection, and including the types of *Pantolambda cavirictus* Cope; *Titanoides zeuxis* Simpson, *Titanoides simpsoni* (described below), *Ignatiolambda barnesi* (described below), *Pantolambdodon fortis* Granger and Gregory.

Dr. L. S. Russell, National Museum of Canada, for a pantodont upper molar from the Saunders Creek beds of Alberta.

Dr. John A. Wilson of the University of Texas, for making available for study two specimens of *Coryphodon* sp. from Eocene beds in Big Bend National Park, Brewster County, Texas.

Thanks are due to Drs. E. H. Colbert, Joseph T. Gregory, and C. L. Gazin, who have also generously given of their time for discussion of some of the basic ecologic and taxonomic problems raised by this group of mammals.

I should also like to thank the staff of the Department of Vertebrate Paleontology of the American Museum of Natural History for assistance in locating various papers and specimens during the course of this study.

Lastly, thanks are expressed to Mr. F. N. Goto, Preparator at Princeton University, for his assistance in preparing the pantodont specimens at Princeton.

ABBREVIATIONS OF INSTITUTIONS

(With reference to their respective collections of vertebrate fossils.)

AMNH	American Museum of Natural History
CM	Carnegie Museum
CNHM	Chicago Natural History Museum
IVP	Institute of Vertebrate Paleontology, Peking, China
KU	University of Kansas
NMC	National Museum of Canada
PIN	Paleontological Museum, Academy of Sciences, USSR
PU	Princeton University
USNM	United States National Museum
UT	University of Texas
UW	University of Wyoming

II. HISTORY OF STUDY

Although recent discoveries have entirely changed earlier conceptions as to the nature of the Pantodonta, some knowledge of this group goes back almost to the beginning of the study of fossil mammals. Cailleux (1945) reports that the first member of this order to be discovered was a skeleton belonging to *Coryphodon* sp. which came to light near Soissons, France, in 1807, but that the specimen was lost before the paleontologist Cuvier learned of its existence. Consequently, not Cuvier but the British anatomist, Sir Richard Owen, was the first to describe a member of the group of mammals now designated as the Pantodonta. This he did in 1845, basing his description of the new form, *Coryphodon eocaenus*, principally on a lower right third molar and the posterior half of the second, from the London clays. Owen initially recognized the tapir-like appearance of these lower molars—certainly a striking example of convergence between the two forms. He reports (1846: 305),

The abundance and variety of the fossil remains of fruits, most of them of a tropical character, which have been obtained from the same deposits of eocene clay as that which yielded the subject of the present section, bespeaks the extent and nature of those dark and dense primeval forests in which the Coryphodon obtained its subsistence. In size, the ancient British Tapiroid quadruped must have surpassed the largest Tapir of South America, or Sumatra, by one-third. The unique fossil specimen which has led to its determination was dredged up from the bottom of the sea, between St. Osyth and Harwich on the Essex coast, and now forms part of the interesting and instructive collection of my esteemed friend, John Brown, Esq., of Stanway Green, near Colchester. The specimen is petrified, and heavily impregnated with metallic salts; it presents the usual rich deep brown colour of the fossil bones of the London clay: the pyritic matter which sparkles in the cancelli of the bone, and which lines the pulp-cavity of the broken molar tooth, leaves no room for doubt as to the fossil having been originally imbedded in that eocene tertiary formation of the Harwich coast.

In 1856 Hébert described a new species of Owen's genus from specimens collected in the Paris basin and in Owen's honor named the new species *Coryphodon oweni*.[1]

During the 1870's discoveries of *Coryphodon* were made in North America by Cope and Marsh, but at first Cope (1872a) did not recognize their similarity to *Coryphodon* and therefore placed these specimens in a new genus *Bathmodon*. In 1873 Marsh pointed out the error, but Cope, although recognizing that some of these specimens belonged to *Coryphodon*, established four other genera as well, and eventually described twenty-one species. Earle in 1892 reduced these to ten species and three genera, and Osborn (1898b) referred all these genera to one genus, *Coryphodon*, recognized ten species previously described, and established three new ones.

Because of the resemblance between the lower molars of *Coryphodon* and of the tapir, Owen regarded this form as a perissodactyl—an opinion echoed by Marsh in 1876 when he proposed a distinct family for this genus, the Coryphodontidae. However, in the previous year Cope had already established for *Coryphodon* a distinct suborder, the Pantodonta, and he placed this group, along with the Dinocerata (which Marsh had proposed as a distinct order in 1872) in a new order, the Amblypoda.

The ensuing taxonomic history of the pantodonts has been summarized by Simpson (1937a: 265) and will not be repeated here, except to point out that the association of the uintatheres (Dinocerata) and the pantodonts has been proven erroneous as a result of the work of Wood (1923), of Simpson (1929 and later), and of Patterson (1939a).

A further clarification resulting from the work of these authors was the demolition of Cope's suborder Taligrada, due to the removal of the Periptychidae by Simpson (1937a: 216–224) and the demonstration of affinity between *Coryphodon* and the pantolambdids brought about largely as the result of the work of Patterson (1939 and earlier).

The recent discoveries included here tend to con-

firm the essential unity of the order Pantodonta as redefined by Simpson (1937a: 265–269), and to emphasize their primarily subungulate affinities.

III. SYSTEMATIC REVISION

INTRODUCTION

Ideas about the taxonomic relationships of the archaic mammalian order Pantodonta, both to other orders and within the group itself, have previously gone through several distinct stages. A highly subjective element will probably always be present in the study of this group of mammals which is so far removed from living forms and which exhibits such a curious admixture of similarities in the posterior dentition with pronounced structural distinctions in the anterior teeth and in the pes. To some students there seems to have been, in the past, an unwarranted amount of splitting of species in this order and the production of too many monotypic genera.

In the classification here attempted, however, an effort has been made to avoid basing taxonomic distinctions on slight characteristics that do not suggest adaptations of the sort that in living forms are associated with discrete populations (or species). It should also be pointed out that for genera such as *Titanoides*, *Barylambda*, *Leptolambda*, and *Pantolambda* there are now known many partial skeletons. This multiplicity of specimens assigned to these genera strongly reinforces their validity, but at the same time other fragmentary materials show that recovery of a reasonably full sampling of the Paleocene pantodont genera and species of North America is by no means complete. The possibility remains that through splitting of a sort not attempted here (based on size differences, for example), additional species could be established. Further discoveries, moreover, will certainly alter the present taxonomic interpretation.

In order to amplify classificatory data for the Pantodonta, the ratio diagrams included in part A of the Appendix were constructed.[2] These diagrams can be informative in two primary regards: first, in representing the distinctions in proportion between the dentitions of different pantodont species, and second, in indicating the variability of proportion among individuals within a single species. It could be argued that not much taxonomic weight should be given to such comparisons of linear proportions and that a more direct and significant analysis is the evaluation of cusp shape, and occurrence or absence of dental structures. For the Pantodonta, however, the differing length-width proportions in the tooth series, particularly those of the lower dentition, seem to reflect rather consistent and taxonomically significant differences. An example

---

[1] Cailleux (1945) suggested that *Coryphodon eocaenus* and *Coryphodon oweni* differ in size and age; the former being smaller and older.

[2] These diagrams are based on the logarithmic scale following the suggestion of Simpson (1941: 23), which is also briefly outlined at the beginning of Appendix A.

of such a difference in proportions is that the upper molars of *Titanoides* consistently increase in size from front to back, while in *Haplolambda* this series typically decreases in size posteriorly.

One possible explanation of the combination of pronounced similarities in the dentition with more markedly different post-cranial structures, mentioned above as being characteristic of the Pantodonta, could be that this is due to our knowledge of the order at a time close to its initial radiation. Since the origin of nearly all Cenozoic orders of mammals appears to have taken place in the Paleocene epoch, or before, and the Pantodonta are the earliest of these groups for which many almost complete skeletons have been recovered, there is presumably no other Tertiary mammalian order that is so well known at such an early stage in the evolution of the placental Mammalia. The Paleocene pantodonts probably represent part of the initial adaptive radiation of the order. In North America at least, there were not many other large mammals in the middle Paleocene. Consequently, the pantodont stock would have been more free to differentiate into a variety of basic adaptive types than has usually been possible in one phylum in later times. For instance, neither the order Carnivora nor Artiodactyla exhibit, during any one Tertiary epoch, so wide a diversity of foot types as do the Pantodonta, which include forms with fully developed claws, flattened fissured unguals, and elephantine unguals. Competition between the various species of large pantodonts themselves might have been the compelling force to bring about selective differentiation of the order. If this hypothesis is correct, it may answer the objection that in the absence of many other large mammals creating competitive pressures, the group would tend to remain undifferentiated. Of course, it would be premature to assume that the full complement of large mid-Paleocene terrestrial vertebrates has yet been discovered. Certainly enough is known to suggest that the diversity of pantodont types arose as the result of the absence of an extensive large mammalian fauna, and that the Tiffanian and later pantodont radiation occurred in a way somewhat analogous to the flowering of groups of mammals isolated from competing forms on islands, such as the Lemuroidea on Madagascar, which has produced such distinctly adapted forms as *Daubentonia* and *Megaladapis*.

The following classification takes into account the observation of Simpson (1953: 342) that "our recognition of a higher category is *ex post facto*" and that in some ways the resemblance between the extant orders increases as one goes back in time so that, as he remarks elsewhere, if a taxonomist of the Mammalia were living in early Tertiary times, he might be tempted to consider many of the primitive groups of mammals only families instead of orders. In other words, part of the justification of the present mammalian taxonomy of the early Cenozoic is the result of

projecting backward in time concepts that are rooted in the more highly distinctive end products of the various mammalian orders living today. Some doubt has been cast by previous writers on whether or not the Paleocene orders in general, and the Pantodonta in particular, are based on as distinctive characters as those which separate some living orders, for example, Chiroptera, Cetacea, and Perissodactyla.

The same problem exists when one is attempting to justify the validity of early Tertiary families within an order, and the degree of distinctiveness which makes necessary familial separation has been a major point of consideration in constructing the classification which follows. The present study has indicated that there are at least four major adaptive types belonging to this order and that there are no species which are certainly transitional between any of these four groups. Three of these assemblages, the Coryphodontidae, Pantolambdidae, and Barylambdidae, have already been recognized as distinct families. The realization that the pantodonts included in the genus *Titanoides* represent a highly specialized stock with digging claws and transversely flattened, sabre-like upper canines has necessitated their removal from the Coryphodontidae as it was last defined by Patterson (1939a: 370), and the resurrection of the Titanoideidae (Scott, 1937b) for their reception. As long as forms transitional between the Pantolambdidae and Coryphodontidae (*Coryphodon, Eudinoceras, Hypercoryphodon*) are not known, it seems preferable to retain the family Pantolambdidae as a distinct primitive stock. Placing the latter family in any one of the three other divisions would be arbitrary, for the Pantolambdidae are about equally removed from them all.

It is unlikely that there was a sufficient lapse of time between the late Torrejonian, in which the earliest pantolambdids appear, and the early Tiffanian from which the first barylambdids are known, for the former to have given rise to the latter. It is still less likely that the Pantolambdidae are ancestral to the Titanoideidae which are known from the Torrejonian, at levels presumably very close in time to the earliest pantolambdids. The phyletic connections of the Coryphodontidae with any of the other three families are entirely uncertain, and the assumption that they are descended from any of the known earlier groups of pantodonts is not indicated by present evidence.

The aberrant Asiatic mammalian family, known only from the Eocene of Inner Mongolia, the Pantolambdodontidae (Granger and Gregory, 1934), are not included in this classification because a reexamination of the lower dentitions of *Pantolambdodon inermis* and *Pantolambdodon fortis* at the AMNH has placed still more doubt on their originally tenuous reference to the Pantodonta. Simpson has suggested (personal communication) that the two species may not even be distinct, and it would appear from the original description that *Pantolambdodon fortis* is distinguished from *Pantolambdodon inermis* by size alone. The latter species is approximately twenty per cent smaller in the few corresponding measurements that can be taken in the two forms.

Granger and Gregory (1934: 5), after a wide ranging discussion of their reasons for not referring this family to any order other than the Pantodonta [Amblypoda], have the following to say about affinities with this group:

When we come to *Pantolambda* and *Titanoides*, however, we find some apparently reliable indications of remote relationship to the Mongolian types, especially in the form of the premolars and molars. In spite of the fact that *Titanoides* is a graviportal form almost as big as *Coryphodon*, it shares the following features with the Mongolian fossils: (1) dental formula of primitive placental type; (2) P_1, P_2 compressed; (3) talonid fossa of premolars formed between the posterior ridge connected with the main cone and a transverse metaconid ridge; (4) molar talonids with V-shaped crests; (5) talonid of M_3 narrower than trigonid; (6) M_3 with reduced or no hypoconulids; (7) M_1, M_2 with no trace of hypoconulid; (8) molars not crowded but slightly spaced; (9) coronoid process inclined backward. *Titanoides* is distinguished from the Mongolian forms by its relatively gigantic size, powerful, more erect incisors and canines; relatively shorter, more massive jaw, etc.[3]

Admittedly the Pantolambdodontidae are a most difficult group to assign, and Granger and Gregory's tentative association of them with *Pantolambda* and *Titanoides* must have seemed at that time preferable to considering the group *incertae sedis*. Further, on negative grounds, it was more reasonable in 1934 to refer this group to the Pantodonta than it is today, because the close similarities of all pantodont lower dentitions, particularly of the lower premolars, were not then fully realized. Consequently, the vague resemblances of the lower dentition of *Pantolambdodon* to the Pantodonta could more easily be supposed to be within the limits of the order. Moreover, it does not appear that the occurrence of the nine characters quoted above in both groups is an adequate indication of relationship. The first and second characters in which the two groups agree are typical of many other orders, for example, Notoungulata and Litopterna. The third character concerning the construction of the talonid fossa of the premolars is not valid, for the dominant posterior ridge of the protoconid of *Pantolambdodon* is joined at right angles by the metaconid crest, while in the Pantodonta the continuous crest runs from the protoconid to the metaconid and is joined at right angles by the posterior ridge, here considered serially homologous with the *crista obliqua*—a distinctly different condition. The fourth, fifth, and seventh characters given by Granger and Gregory constitute real similarities between the two groups; but their many dissimilarities strongly suggest, if they do not confirm

(in the minds of some workers) that these three resemblances do not reflect close affinities. The sixth proposed feature of resemblance, reduced or absent hypoconulids on the M_3, is not very relevant because hypoconulid cusps do occur here in *Coryphodon* and *Titanoides*. The eighth feature is incorrect because pantodont molars are typically crowded. The ninth character is that the coronoid process was said to be inclined backward in a similar way in both groups, but no known pantodont species has such a backward inclination of the coronoid process as that present in *Pantolambdodon*.

Granger and Gregory (1934: 6) add that: "From *Pantolambda* the present form differs in its much more elongate slender jaw, somewhat procumbent front teeth, more hypsodont cheek teeth, compressed premolars; the molars have much larger anterior V's; the ascending ramus slopes backward and is distinctly delicate." To these differences may be added the following distinctions which appear to be adequate to remove *Pantolambdodon* from the Pantodonta entirely. First, Dr. G. L. Jepsen on examination of the two paratype specimens of *Pantolambdodon inermis* AMNH 22100 and 21748, collected in the same year from the same locality[4] has recently discovered that on the posterior surface of the supposed M_3 of AMNH 22100 there is a mitten-shaped wear surface which fits exactly into an identical wear surface on the anterior face of the supposedly isolated M_3, AMNH 21748 (see fig. 17). The separate molar also continues the molar series of AMNH 22100, for in this specimen there is a progressive reduction from M_1 to M_3 of the size of the talonids, coupled with an increasingly wider angle between the anterior and posterior talonid crests from front to back; and the isolated molar follows both of these progressions. The conclusion that AMNH 21748 belongs with the other paratype as a lower M_4 seems inescapable. To those acquainted with mammalian taxonomy, this would appear to be a rash conclusion, but x-ray photographs and extensive comparisons with all specimens of *Pantolambdodon* make the presence of an M_4 in this genus highly probable. There are, however, two other possible interpretations of this sequence of four molariform teeth in the paratype of *Pantolambdodon*, AMNH 22100. One is that the first tooth in the series is a molariform milk P_4. In this case there should almost certainly be some indication of a permanent P_4 beneath the tooth, but x-rays do not reveal any such structure. Another alternative is that there was a molariform permanent P_4 which, if actually the case, is an entirely non-pantodont feature. Moreover, comparison of the type and paratype of *Pantolambdodon inermis* strongly suggests

---

[3] Evidently Granger and Gregory, when they speak of *Titanoides*, are referring to *Barylambda faberi*, which was originally assigned to *Titanoides*.

[4] This locality is given on the specimen cards as four miles north of Tukhum Lamasery, Shara Murun region, Inner Mongolia, and in Granger and Gregory (1934) as eight miles north of Tukhum Lamasery, but there is no reasonable doubt that the two specimens were collected at the same locality.

that there were four premolars, a canine, and three incisors anterior to the four molariform teeth. If the latter interpretation is correct, it rules out the second and third possibilities mentioned above.

The type of *Pantolambdodon inermis* AMNH 21558 may represent a young individual, judging by the lesser degree of wear on the M_{1-2} and the extreme shallowness of the horizontal ramus (twenty per cent shallower than that of the paratype). Unfortunately, the type specimen lacks the M_3, but posterior to the alveolus of the M_3 is a narrow cavity similar to that which often forms above a molar bud before it erupts. Although x-rays do not indicate a fourth molar unerupted, the presence of this cavity seems to be a further indication of a M_4 in this genus, particularly when this feature is taken together with the evidence of the paratype specimen. Another, but more implausible interpretation would be that there were more than four premolars in *Pantolambdodon*.

Reference of *Pantolambdodon* to the Marsupialia might be considered, but as far as is known at present no large marsupials existed anywhere in Asia during the Eocene epoch.

The following differences of the Pantolambdodontidae from the Pantodonta should be added to those differences noted by Granger and Gregory: condyle of jaw much higher above the tooth row, jaw from base of ramus to condyle rounded with no posterior or ventral extension of the angle, and external division of molar trigonid and talonid shifted more posteriorly so that the groove between the two parts extends down the labial side of the posterior root.

It is to be hoped that the upper dentition of *Pantolambdodon* may eventually be recovered, but until more complete evidence of this sort is found, there is essentially no basis for considering this genus a member of the Pantodonta. The case for including "*Archaeolambda*" Flerow (1952) in this order is sound, as will be pointed out in the discussion of the Barylambdidae. I would suggest that the Pantolamdodontidae be considered protungulates, *incertae sedis.*[4a]

ORDER **PANTODONTA** COPE, 1873
(as suborder)

Diagnosis: Archaic subungulates, middle-sized (size of a fox, *Vulpes*), to large (size of a rhinoceros, *Diceros*). In all species the dentition is typically unreduced (I 3/3, C 1/1, P 4/4, M 3/3), and brachyodont, showing some lophodonty. Upper molars showing primitive para- and metacone W-shaped crests, or retaining some indication of this condition (DP[4] of *Coryphodon*). Upper premolars with strongly developed double V-shaped crests of paracone, and parametacone. Canines varying in size from very small to

[4a] Recently Dr. McKenna has pointed out to me a number of interesting similarities between *Pantolambdodon* and certain Hyracoidea.

tusklike. Skull and jaws comparatively massive with powerful muscle attachments, but usually small compared to body size. Brain small and primitive, and showing little advance through time or with increase in body size. Limbs massive and strong, and typically retaining primitive features such as free (or distinguishable) centrale, separate radius and ulna, tibia and fibula, and with five-toed foot. Alternating arrangement of pes retained. Distribution: North America, from middle Paleocene through early Eocene. Eurasia, from early Eocene (? late Paleocene) through middle Oligocene.

SUPERFAMILY **CORYPHODONTOIDEA** (new rank, definition as for included family)

FAMILY **CORYPHODONTIDAE** MARSH, 1876
(new sense)

Type: *Coryphodon* Owen, 1845, Odontography, or a treatise on the comparative anatomy of the teeth, etc., p. 607.

Included genera: *Coryphodon, Eudinoceras, Hypercoryphodon.*

Distribution: Paleocene, Clarkforkian stage, Wyoming to Eocene, latest Wasatch stage, North America. Early Eocene, Sparnacian and equivalent stages, Europe. Late Eocene of Western Mongolia, Ichang (Hupeh), China, and Dzungaria Basin, Sinkiang: Middle (?) Eocene, Kaojichuan, Sintai district, Shantung, China: to Middle Oligocene of Inner Mongolia, Asia.

Diagnosis of family: Large, specialized, sub-graviportal forms, with full placental dentition. Incisors, above and below, enlarged and spatulate; directed somewhat anteriorly. Canines above and below enlarged and tusklike and usually angled outward as part of a procumbent arc including the incisors, giving the muzzle a hippotamoid flare; lower canine in some forms having an anterointernal ridge running upward from base, sometimes with an analogue of the premolar talonid on the posterior base of the lower canine. Diastemata, above and below, between canine and first premolar. Paraconids of lower premolars and molars, small; *crista obliqua* of lower molars reduced or absent. P_1 either one- or two-rooted, usually having two fused roots; protocone distinct. P^{2-4} with labial tips of parastyle and metastyle more compressed anteroposteriorly

---

FIG. 1. Lateral views of pantodont skulls.
All approximately × 1/5.

A. *Coryphodon* Owen.
B. *Barylambda faberi* Patterson.
C. *Titanoides primaevus* Gidley.
D. *Haplolambda quinni* Patterson.
E. *Caenolambda* Gazin. Mandible *Caenolambda jepseni*, cranium *Caenolambda pattersoni*.
F. *Pantolambda cavirictus* Cope.
G. *Pantolambda bathmodon* Cope.

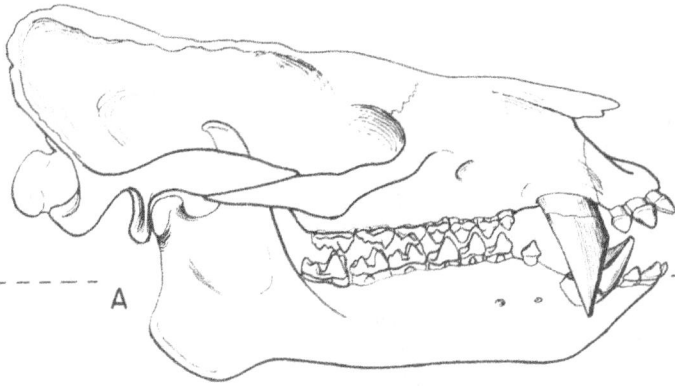

WASATCHIAN

CLARKFORKIAN

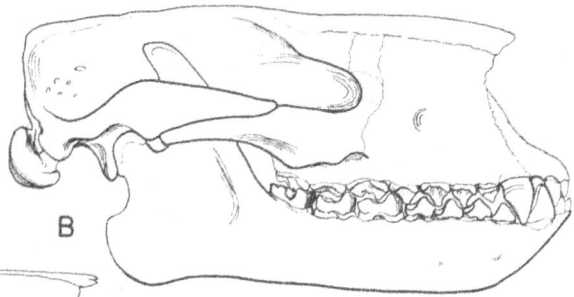

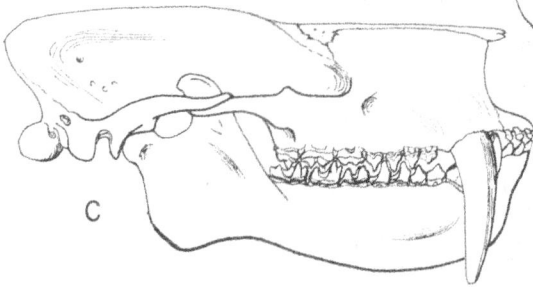

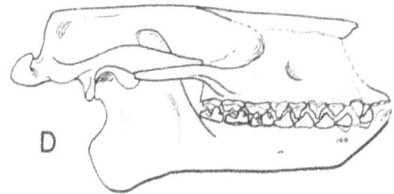

TIFFANIAN

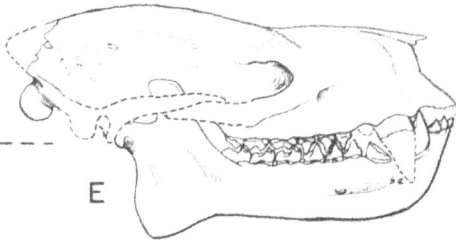

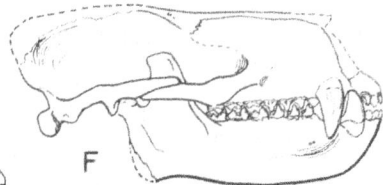

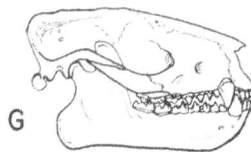

TORREJONIAN

than in other families. Upper molars quadrate or sub-circular, never triangular; primitive W of paracone and metacone crests modified into a bilophodont pattern.

Nares terminal anteriorly, moderately large; nasals long, of medium width; premaxillaries large and heavy, with slight facial extension of ascending rami. Facial region only slightly deeper than cranial and with pronounced fossa behind the upper canine. Posterior roof of skull broad and flattened. No postorbital processes. Ventral surface of palate broadly expanded in region of internal nares, particularly in the Mongolian species. Post-glenoid process comparatively large and relatively close to mastoid process; paraoccipital process not entirely distinct from mastoid process. No broad exposure of anterointernal ventral face of petrosal on basicranium, as in Titanoideidae.

The post-cranial skeleton of the Coryphodontidae as here defined is known only in *Coryphodon;* its description is consequently given under that genus.

Discussion: The Coryphodontidae are by far the longest surviving and successfully adapted family of the order. In this group specialization of the dentition reaches its most advanced stages. There is evidently a trend in the dentition, from basically dilambdodont molars, present in the primitive species of *Coryphodon* toward the lophodont or sub-lophodont dentition of *Hypercoryphodon* of the mid-Oligocene Houldjin Gravels of Inner Mongolia.

I should like to thank Dr. Malcolm McKenna for pointing out that Flerow's new pantodont species *"Procoryphodon primaevus"* might be *Phenacolophus fallax* (Matthew and Granger, 1925: 9) which is also from the Gashato formation of Mongolia. The type of *Phenacolophus,* AMNH 20411, as well as that of *"Procoryphodon,"* is rather poorly preserved, but, as the two forms agree in all clearly distinguishable features, there is no basis for separating them. *"Procoryphodon"* therefore falls into synonymy with *Phenacolophus.* It should also be pointed out that the supposedly primitive features of *"Procoryphodon"* put forward by Flerow as relating it to the ancestry of *Coryphodon* are not supported by the lower molar structure seen in the late Paleocene and early Eocene species of the latter genus from North America. At present the taxonomic placement of *Phenacolophus* is, Condylarthra, *incertae sedis.* The upper molars and lower premolars of this form show that it has nothing to do with the Pantodonta. In fairness to Flerow it should perhaps be mentioned that the bilophodont M_3 of *Phenacolophus* is rather like that of *Coryphodon* but the resemblance is no closer than that seen in the lower molars of some marsupials and in the tapir.

In the most structurally primitive dentitions of the Coryphodontidae, such as that of *Coryphodon proterus* of the Clark Fork Paleocene (diagnosed below) and *Coryphodon cinctus* of the Wasatch Eocene, reduction of the anterior crests of the trigonid and talonid V's of the lower molars has progressed over that of the earlier pantodonts.[5] Other Wasatch species such as *Coryphodon simus* and *Coryphodon armatus*[6] show a further advance in that the dominant lophs run from the protoconid to metaconid, and from entoconid to hypoconid, while the paraconid crest and *crista obliqua* are low and poorly developed. This condition reaches its maximum development in the more bilophodont upper and lower dentition of the Eocene *Eudinoceras* and Oligocene *Hypercoryphodon,* both of which occur in Inner Mongolia.

Eudinoceras mongoliensis was described by Osborn in 1924 from a single third or fourth left upper premolar, AMNH 20101, and as the name would indicate, its primarily coryphodontoid features were not evident at that time. The anterior and posterior crests of the premolar paracone are lacking in *Eudinoceras,* a condition which might suggest a trend toward the virtual absence of this cone in the Dinocerata and consequently an affinity to the latter group. After the discovery of nearly complete upper and lower dentitions of *Eudinoceras mongoliensis* reported by Osborn and Granger in 1931 and 1934, it became evident that the species *Eudinoceras mongoliensis,* as well as a geologically earlier form *Eudinoceras kholobolchiensis,*[7] had their nearest ally in *Coryphodon* and not among the uintatheres. The phyletic relationship of *Eudinoceras* to *Coryphodon* nevertheless remains uncertain. The occurrence of members of the latter genus in Europe, Asia, and North America in the early Eocene at least makes possible the derivation of *Eudinoceras* from *Coryphodon* from the zoogeographic standpoint. Since *Eudinoceras* does not possess any dental features not foreshadowed in the teeth of *Coryphodon,* it could have arisen from some species of the earlier genus. One slight distinction in the two forms is that the upper first premolar in *Eudinoceras* is two-rooted, which is not often the case in the North American species of *Coryphodon.* However, Cailleux (1945) reports that the first upper premolar of *Coryphodon eocaenus* is normally two-rooted, and even in the North American forms the root, if single, is usually separated into anterior and posterior halves by vertical grooves on its labial and lingual sides, suggesting that it may be secondarily one-rooted. No pantolambdid or barylambdid has a two-rooted upper or lower first premolar. Since two-rooted P1's do occur in *Coryphodon,* there is no clearcut distinction from *Eudinoceras* in this respect.

One rare feature of the upper dentition of *Eudinoceras* which Osborn noticed in the type specimen, namely the absence of the premolar protocone crests,

---

[5] This reduction is more typical of specimens of *Coryphodon* of Wind River age.

[6] See Osborn, fig. 16 (1898a: 192).

[7] This species has also been reported from the red beds of Ichang, China by de Chardin and Young (1936).

is partially developed in only two other types of pantodonts, the species of *Titanoides* and in *Haplolambda planicanina*. The latter species is Asiatic, having been recently discovered (Flerow, 1952), in lower Eocene deposits of the Nemegetou Basin, southern Gobi Desert of Mongolia. In spite of the reduction of the upper premolar protocone crests shared by these forms, there is little similarity in detail in this reduction and it apparently does not indicate a close affinity; for in other ways the dentitions of *Eudinoceras*, *Titanoides*, and *Haplolambda planicanina* are quite distinct. The upper premolars of *Eudinoceras* resemble those of *Coryphodon* in that their anterior and posterior margins are convex, while in *Titanoides* they are concave. Most probably, therefore, the reduction of the premolar protocone crests in these forms is due to parallelism. Both *Eudinoceras* and the Oligocene *Hypercoryphodon*,[8] which is known only from a single skull lacking the mandible, share with *Coryphodon* cranial and dental features such as the trend toward bilophodonty, the flaring muzzle, enlarged incisors, anteroposteriorly shortened basicranium and flattened upper surface of the cranium. Presumably these pantodonts are all part of a distinct Asiatic radiation of the coryphodontids, of which *Coryphodon* was the primary, more widely distributed stage. That *Hypercoryphodon thomsoni* was not a direct descendent of *Coryphodon* might be suggested by the much less extensive development of the flattened expansion of the saggital crest present in the type of the genus; but this gracility of the crest (as in the type of *Coryphodon wortmani* Osborn, 1898a: 212) may be neither a primitive nor a specific character, but the result of sexual dimorphism, disease, or of immaturity. In the latter cases, the narrow skull roof of *Hypercoryphodon* would not be a significant distinction from that of *Coryphodon*.

Chow, 1957, has described a new species, *Coryphodon flerowi*, from beds which may be of middle Eocene (Lutetian) age from near Kaojichuan, Sintai district, Shantung, China, but there is little basis for dating this find. This new species is based on an isolated second upper molar, IVP 927, that has been stated to be more progressive than all other known species of the genus. In *Coryphodon flerowi* the posterior protocone crest dies out at the anterointernal base of the metacone instead of joining the posterior cingulum of the tooth as it does in other species of *Coryphodon*—a feature which does foreshadow the virtually complete supression of this crest in *Eudinoceras* and *Hypercoryphodon*. In most North American specimens of *Coryphodon* sp. of Wind River age, such as PU 16824, the reduction of the posterior protocone crest and its separation from the posterior cingulum in the upper second molar are also evident but the condition is not as well developed

as in *Coryphodon flerowi*. However, many *Coryphodon* dentitions of Wind River age do show a somewhat greater reduction of the paracone in the upper second molar than is to be seen in *Coryphodon flerowi*, and in this respect approximate more closely the transverse upper molar "metaloph" of *Eudinoceras* and *Hypercoryphodon*.

CORYPHODON Owen, 1845: 607

Type: *Coryphodon eocaenus* Owen, 1845.
Included species: See Earle (1892: 155) and Osborn (1898: 191, 197).[9]
Distribution: Clarkforkian stage, Paleocene, to latest Wasatch stage, Eocene, North America. Sparnacian stage, early Eocene, Europe. Middle (?) Eocene, Shantung, China.[9a]
Description and Diagnosis: Past studies of this pantodont have been largely concerned with describing (or with questions about the validity of) the twenty-eight proposed species of *Coryphodon*, and little attention has been given to compilation of the characters of the dentition shared by members of this genus. Most of the skeletal distinctions are treated in Patterson's useful paper (1939b) on *Coryphodon ?elephantopus*. For an early definition of this genus, see Earle (1892: 155).

DENTITION

Incisors above and below much larger comparatively than in other pantodont families, with lateral crest on each side giving the tooth a spatulate form, procumbent, and with basal cingulum in some species. Canines above and below large and tusklike, oval in cross section in most species, but not as elongated from front to back at the base as in *Titanoides*. P_1 separated from adjacent teeth by broad diastema anteriorly, and in most species a somewhat smaller one posteriorly. Paraconids of lower premolars and molars reduced. Trigonids of molars higher than talonids, "crista obliqua" (the anterior wing of the posterior molar V) reduced and low, not reaching tip of metaconid. M_{1-2} talonids often wider than trigonids. M_3 talonid in earliest species with distinct entoconid, hypoconid, and hypoconulid cusps; but in structurally more advanced species with entoconid absent and with hypoconulid and hypoconid connected and forming a cross-loph perpendicular to long axis of the tooth and parallel to a

---

[8] The upper incisors of *Hypercoryphodon* reverse the size progression of the earlier pantodonts, in that the I^1 is the largest, judging from the alveoli preserved in the type, AMNH 26384; see Osborn and Granger (1932: 5).

[9] The extreme confusion as to what constitutes a species in this genus makes it futile to list them here. A revision of this group should be facilitated by a collection of parts of fourteen individuals, including complete skeletons of *Coryphodon* from one locality made in 1950 and 1951 by field parties of the American Museum of Natural History in the San José formation of the San Juan Basin of New Mexico.

[9a] Since the above was prepared Chow has reported *Coryphodon* of middle or early Eocene age as occurring in Sinyu, Kiangsi, South China (*Vertebrata Palasiatica* 3 (2), 1959).

similar protoconid-metaconid loph.[10] Lower molars and premolars increasing in size from front to back. P^1 usually exhibiting slight development of a protocone. P^{2-4} somewhat heart-shaped, having convex anterior and posterior margins, and lacking the constriction of the middle of the premolars seen in earlier pantodont genera which gives them concave anterior and posterior margins; parametacone V and protocone V, typical, but parastyle and metastyle directed slightly toward each other at the tips, instead of diverging as in the Pantolambdoidea. M^{1-3} with small paracone, sometimes barely distinct, but with strongly developed crest or protoloph incorporating the protocone, protoconule, and parastyle, which is no longer attached to protocone. Mesostyle and posterior wing of metacone V not always present. Paracone and metacone tending to form a straight crest or ectoloph, so that M^{1-3} are basically bilophodont. Hypocone occasionally present on M^{2-3}.

MANDIBLE

Massive, with least depth posterior to M_3; condyle relatively broad transversely; coronoid process and condyle high, angle usually not projecting as far posteriorly as in Barylambdidae; mental foramena typically situated below P_1 and P_2, occasionally below M_3. Fusion of symphysis complete before eruption of M_3.

CRANIUM

Flattened roof of varying width. Zygomatic arches flaring but not always massive. Muzzle, as in all the Coryphodontidae, flaring; nasal opening not as large as in Barylambda or Titanoides and differing from latter genera in having a distinct facial extension of ascending process of premaxilla. Palate flat with major foramena placed slightly internal to M^1 and with occasional smaller foramena posteriorly. Basicranium with elements more crowded than in earlier genera owing to more posterior position of post-glenoid processes. Condylar foramen present.

VERTEBRAE

Vertebral formula C7, D17, L4, C15. All vertebral centra slightly biconcave. Cervicals relatively large for a pantodont, particularly atlas and axis, which are about twenty-five per cent broader than in Barylambda; relative length of centrum of axis intermediate between Barylambda (shorter) and Titanoides (longer). Thoracics with rather short neural spines, more pronounced ventral keel and longer than in Barylambdidae. Usually four sacrals. Anterior caudals forming a unique elongate pseudosacrum by coalescence of the centra, arches and neural spines; tail reduced in

size and length, chevron bones apparently lacking. Anterior ribs shortest, and flattened at extremities; more posterior ribs oval in cross section, not particularly distinctive.

FORE LIMBS AND STERNUM

Clavicle present, comparatively slender, larger at sternal end. Sternal elements not unlike those of Pantolambda in proportions but with slightly greater width than in other pantodonts. Scapula with suprascapular border forming a somewhat pointed knob above the end of the spine; both acromion and coracoid processes smaller than in Titanoideidae. Deltoid crest of humerus moderately developed, not as high as in Titanoides and Barylambda. Head and distal extremity of radius similar to those in other pantodonts in general proportions, but shaft relatively shorter and lacking the high ridge running down the distal half of the anterior face which is present in Leptolambda. Shaft of ulna distal to greater sigmoid cavity, shorter than in pantolambdoids.

MANUS

Centrale almost entirely absorbed by fusion with scaphoid, lunar usually much more flattened than in earlier genera; lateral extremity of unciform relatively large; metacarpals shortened and relatively broad; pisiform relatively larger than in Titanoides and with more distinct neck than in Barylambda and Leptolambda: proximal phalanges about as long as in Barylambda, relatively longer than in Leptolambda; both proximal and mesial phalanges relatively shorter than in Pantolambda and Titanoides; unguals broad and flattened with expanded distal margin lacking central fissure of barylambdids and pantolambdids; altogether a much more graviportal foot than in other pantodonts, convergent with the manus of the Dinocerata, Astrapotheria, and Proboscidea.

PELVIS AND HIND LIMB

Ilium with greatly expanded anterior border, anteroexternal angle directed slightly backward, but lacking enlargement and thickening of tip of anteroexternal angle present in Barylambda; pubis slenderer than ischium which has relatively broad syphysis. Femur with third trochanter, but not particularly distinctive; pit for ligamentum teres present, patella and trochlea elongate; distal tip of patella much more elongated than in Barylambdidae. Tibia usually with low cnemial crest; fibula articulating with astragalus.

PES

Broad with foreshortened phalanges; astragalar foramen present; calcanium lacking accessory internal tubercle of Barylambdidae; phalanges as in manus, but relatively smaller.

Discussion: The validity of many of the twenty-eight proposed species of this genus has long been questioned.

---

[10] The lophodonty of the lower molars of some individuals of *Coryphodon*, for example PU 13472, is as great as that exhibited by the most advanced known coryphodontid lower dentition, that of *Eudinoceras mongoliensis* (AMNH 26611). The paraconids of PU 13472, however, are not quite as reduced as in *Eudinoceras*.

Since most of the types of the various species described by Cope and Osborn are partial upper or lower dentitions they cannot always be compared. Osborn (1898b: 189) remarks:

Twenty-one species were named by Cope, with as little regard for the laws of individual variation as for the association of skeletons with teeth or of jaws with skulls. It is *a priori* improbable that such numerous species should have coexisted, considering that all the collections come from a few levels and a single geographical region. Our knowledge of large living quadrupeds, such as the African Rhinoceros, shows that rarely more than two species of one genus coexist, and these have different local feeding habits. The writer has found the same to be true of the Eocene Titanotheres of Wyoming. Earle's revision of the species (1892), therefore, marked a valuable advance but left much to be done, owing to his lack of comparative material at the time.

It seems clear also that the *Coryphodon* group was characterized by high individual variability, which is in part responsible for the previous splitting of species. Until large samples of this mammal from the same locality, and presumably of one population, have been analyzed, the most valid specific characters will not be known. For this reason the early revisions of the genus by Earle (1892) and Osborn (1898b) were not entirely satisfactory. A more recent study of the European members of this genus by Cailleux (1945) has suggested the specific distinctness of *Coryphodon eocaenus* Owen, and *Coryphodon oweni* Hébert. Cailleux tentatively considers the European species to be closest to the North American *Coryphodon obliquus*.

The present study indicates that, much more than in the taxonomy of other pantodonts, most of the species of *Coryphodon* are based on characters which probably have little functional significance and which consequently do not suggest distinctly adapted species. It is therefore greatly to be desired that the large undescribed collections of *Coryphodon*, particularly those at the American Museum of Natural History, the Chicago Natural History Museum, and the U. S. National Museum, be studied in order to clarify the status of the North American members of this genus.[10a] Pantodont species, in general, do not appear to have had long temporal ranges; and it is likely that, if the taxonomy of the species of *Coryphodon* can be revised, the various species will serve as good index fossils for detailed stratigraphic comparisons of beds of early Tertiary age. The large size of these pantodonts increases the likelihood of preservation and discovery, so that they fulfill as well a second requirement of a good index fossil, ubiquity. The volume of known material of various species of this genus is so great that a taxonomic revision would be an undertaking of several years' duration and consequently cannot be reported here. Nevertheless, some consideration of the major trends in the evolution of *Coryphodon* is necessary in connec-

tion with this study for two major reasons: First, comparisons between this genus and the earlier pantodonts are essential for the reconstruction of pantodont phylogeny. Second, the Paleocene specimens in the Princeton collections, described below, reveal some interesting primitive characteristics of species of this genus which should be pointed out. Data derived from comparative studies of Eocene forms of *Coryphodon* appear in this paper in the descriptions of the other types of pantodonts as well as in Section IV on anatomy and in the ratio diagrams included in Appendix A.[11]

Coryphodon proterus[12] sp. nov.

Figure 18

Coryphodon sp. indet. Jepsen, 1930b, *Proc. Amer. Philos. Soc.* 69 (7): 493.
Coryphodon sp. nov. Van Houten, 1945, *Journal of Paleontology* 19 (5): 425.

Type: PU No. 13400. Left lower jaw with C, P_2-M_3, left and right I_{1-3}, right P_{1-4}, M_1.

Horizon and locality: Clark Fork, Paleocene; three miles SW of Bear Creek, Carbon County, Montana.

Specific diagnosis: Incisors comparatively smaller, particularly in transverse diameter, than in later species of *Coryphodon*. Lower canine very large, broad transversely, bearing pronounced anterointernal wing or ridge. Diastema posterior to P_1 short. P_{2-4} resemble *Coryphodon testis* sort, but are somewhat more quadrate at base, with wider angle between crests from protoconid to paraconid and metaconid than is usual in other species; base of paraconid strongly produced anteriorly. M_{1-3} distinctive in that the part of the talonid internal to the *crista obliqua* makes up a flattened horizontal plane which is higher than the external part of the trigonid. In other types internal and external portions of the talonid usually drop away from *crista obliqua*. Slight metastylids present on M_{1-3}.[18] M_3 large relative to M_{1-2} with three quite distinct talonid lobes: hypoconid, hypoconulid, and entoconid. The species appears to be distinctive in possessing a crest from hypoconulid to entoconid equal in length to that from hypoconid to hypoconulid, and in having grooves descending along the sides of the posterior root at the base of hypoconulid of M_3 which separate off a posterior lobe of the tooth. (For all measurements of dentitions see Appendix B.)

Discussion: In figure 8 this specimen is compared with twenty lower dentitions of Eocene species of *Coryphodon*, principally from the Bighorn Basin of

---

[10a] The writer is currently preparing a taxonomic revision of the species of *Coryphodon*.

[11] For further discussions of the characters of *Coryphodon* see Cope (1872, 1874, 1875, etc.), Marsh (1876, 1877, 1893), Osborn (1898a), and Patterson (1939b).

[12] From the Greek πρότεροι, earlier, with reference to the Paleocene age of the species.

[18] The presence of molar metastylids in *Coryphodon* is not particularly common, as pointed out by Patterson (1939b: 100). They are absent in eighty per cent of the specimens examined for this study.

Wyoming. The type of *Coryphodon proterus* in most dimensions is well above the average size of these specimens, which may represent several species. *Coryphodon proterus* in a few measurements is above the extremes of range of the other individuals, although with a larger comparative sample this almost surely would not be true. In some respects, particularly the strong development of three talonid cusps, the lower molars of *Coryphodon proterus* are most like those of AMNH 4329 which has been referred to *Coryphodon cinctus*.

The most internal cusp of the lower M_3 talonid of *Coryphodon* has been called the entoconid 2 (Osborn) or the entostylid. In *Coryphodon proterus*, however, it seems most reasonable to consider that entoconid, hypoconulid, and hypoconid are all present, and that such a condition represents a more primitive stage of M_3 conformation in this genus, closely resembling the M_3 talonid of *Pantolambda bathmodon*. In most Eocene species of *Coryphodon* the hypoconulid and entoconid, or entoconid 1 and 2, have been reduced from the presumably primitive condition of *Coryphodon proterus* into a straight crest or spur running internally from the hypoconid. As far as M_3 is concerned the cusp which is lost is the most internal (entoconid).

Clearly Earle (1892: 154) was mistaken in his conjecture that "The most primitive condition of the last inferior molar in *Coryphodon* is probably where the heel has a straight posterior border, for this is the condition in *Pantolambda*." Actually, the heel of the M_3 of *Pantolambda* does *not* have a straight posterior border but possesses instead a well-defined cusp, posterior to the hypoconid, which is situated above a distinct lobe of the posterior root as in *Coryphodon proterus*. Consequently, the three-cusped third molar talonid is not an advanced feature as Earle thought but is evidently the primitive condition for all pantodonts, including the *Coryphodon* stock. None of the Eocene species of *Coryphodon* shows quite as distinct an entoconid and hypoconulid as does *Coryphodon proterus*, but even before this species was known, Osborn (1898b: 192) had suggested, in variance with Earle's statement quoted above, that the tri-cusped M_3 talonid is the primitive condition in *Coryphodon*. PU 13400 appears to confirm Osborn's suggestion. The best known Eocene species of *Coryphodon* form two groups with regard to this character. Such species as *Coryphodon eocaenus*, *Coryphodon testis*, and *Coryphodon cinctus* retain the three M_3 talonid cusps more or less distinctly, with the hypolophid set at an oblique angle to the metalophid, while in *Coryphodon simus*, *Coryphodon armatus* and specimens of Wind River age, the metalophid and hypolophid are parallel, creating a tapiroid, bilophodont tooth similar to that of *Eudinoceras mongoliensis*. When the hypoconulid and entoconid are present in the later and presumably more progressive Wind River forms, they are usually on an almost straight line with the hypoconid, not arranged as in

Coryphodon testis and *Coryphodon proterus* at the corners of an almost equilateral triangle.

Although it may be somewhat unsatisfactory to describe another species of *Coryphodon* until a revision of the genus has been completed, the lower dentition of *Coryphodon proterus* appears to be more distinctive than that of any other species of this genus. It is regrettable that no well-preserved upper dentition of *Coryphodon* of Clarkforkian age is known, because the uniquely primitive features of the lower teeth of *Coryphodon proterus* suggest that an upper dentition of this age might more clearly indicate from which of the earlier pantodont stocks the Coryphodontidae are derived. The partial upper and lower dentition from Clark Fork beds mentioned by Simpson (1937c: 23), AMNH 16078, collected in the Clark Fork Basin in 1912 between Little Rocky and Line Creeks in northwestern Wyoming is too poorly preserved to indicate its specific affinities; but the M_3 is large and three lobed.

It is not clear whether *Coryphodon* replaced such Tiffanian genera as *Titanoides* and *Barylambda* through selective competition, or whether the appearance of *Coryphodon* in North America was subsequent to the extinction of the Tiffanian pantodonts. *Coryphodon* has never been reported from Tiffanian levels, notwithstanding Matthew's remark to the contrary (1937: 180). Most of the known Tiffanian pantodonts can be assigned to the lower part of that stage. This limitation may be more apparent than real, however—the result of inadequate collecting in upper Tiffanian levels or the absence there of the faunal facies in which the Pantodonta occur. *Coryphodon* has not yet been collected at the same locality or level with any other pantodont.

Coryphodon, sp. indet.

Figure 16

Two partial skeletons of *Coryphodon*, PU 14682 and 14685 from the Reiss locality, Sec. 20, T 55 N, R 98 W, Bighorn Basin, Wyoming, also appear to be of Clarkforkian age or earlier. The stratigraphic position between the Cretaceous exposures to the northwest and the Willwood Eocene to the southeast, as well as the general appearance of the beds, drab gray and buff-colored shale, suggests that these rocks are of Paleocene age. However, the associated fauna is not complete enough to make exact faunal correlations. Mammals from this locality include fragmentary remains of:

> *Probathyopsis* sp.
> *Psittacotherium?* sp.
> *Ectocion* sp.

None of these specimens is complete enough to indicate whether they are Paleocene or Eocene species. However, at approximately the same level as the *Corypho-*

don skeletal remains a small collection of fossil plants including the following species has been made:

> *Quercus praegroenlandica* Berry
> *Cercidiphyllum arcticum* Newberry
> *Ginkgo adiantoides* Heer
> *Sapindus obtusifolius* Lesquereux
> *Elaeodendron serrulatum* Ward
> *Alnus serrata* Newberry

Dr. Erling Dorf of Princeton University reports that "this is definitely a Paleocene assemblage, but its position within the Paleocene cannot be determined without additional collecting to secure a larger flora."

Most probably then these two skeletons of *Coryphodon* are of Clarkforkian age, if not earlier. Consequently, they constitute the oldest post-cranial remains of this genus, but comparisons with Eocene skeletons of *Coryphodon* have failed to disclose any clearly distinctive features in them.

Unfortunately, the only part of the dentition associated with either of these specimens is a fragment of the premaxilla containing what appear to be parts of erupting incisors. Consequently, specific identification is not possible. This fragment belongs to PU No. 14685, which is the more complete skeleton. The following description is of this specimen: Five cervical vertebrae are present including part of the atlas. The other cervicals are probably C3–6. These vertebrae are fragmentary, but in their large size in proportion to the rest of the skeleton, and in all other observed features, they do not differ from *Coryphodon*. There are twelve thoracic vertebrae which are relatively well preserved and in which the height of the spine resembles closely those figured by Osborn (1898b: fig. 23). There is one thoracic vertebra which resembles the antepenultimate vertebra of *Coryphodon*. Four sacral vertebrae are preserved, attached to the pelvis and posterior to them three caudal vertebrae. The centra and transverse processes of these caudals are fused to each other and to the posterior sacrals. The neural spines of the three posterior sacrals and the three anterior vertebrae are fused into one continuous ridge, in a manner which, according to Patterson (1939b: 103) appears to be typical of *Coryphodon*. No such pseudosacrum is known to occur in any of the Torrejonian and Tiffanian pantodonts.

The scapula of this specimen also resembles that of Eocene *Coryphodon* sp. in the preserved parts, but unfortunately the apex, which is diagnostic of *Coryphodon*, is missing. The subscapular surface agrees in conformation with scapulae of the latter genus, at the American Museum of Natural History and elsewhere, in being somewhat concave. The acromion process is more distinctive, being broad and flattened. A complete, well-preserved pelvis that also resembles *Coryphodon* pelves in all its features is preserved with this specimen. It is illustrated here in figure 16.

The femur, patella, and tibia of PU 14685 also agree with these bones in *Coryphodon* in their general conformation, as do several metapodials and a single proximal phalange.

The less well-preserved specimen PU 14682 includes a fragmentary and much distorted femur, as well as tibia, patella, calcanium, and astragalus of the right side. These bones also agree with *Coryphodon* in size and proportion. The fibula appears to have a distal articular surface with the calcanium and astragalus and in this feature it differs from *Barylambda* and agrees with *Coryphodon*.

SUPERFAMILY **PANTOLAMBDOIDEA** (new rank)

Diagnosis: Middle to late Paleocene forms (possibly early Eocene); ambulatory or sub-graviportal; size medium to large; retaining digits of sub-equal size. Differ from Coryphodontoidea in having comparatively smaller brains and skulls, and with forelimbs indicating great mobility, large clavicle, pelvis and hind limbs usually much larger than forequarters, anterior caudals not fused to sacrum, centrale of manus separate (or forming distinct lobe when fused to scaphoid), at least some unguals with central fissure, tail large and long; upper molars with primitive paracone and metacone forming W-shaped crests, showing no approach to bilophodont condition (as seen in *Eudinoceras*), typically no diastema behind lower canine, incisors small.

FAMILY **PANTOLAMBDIDAE**, COPE, 1883

Type: *Pantolambda*, Cope, E. D., 1882, *Amer. Nat.* 16: 418.

Included genera: *Pantolambda, Caenolambda*.

Distribution: Late Torrejonian (Upper or *Pantolambda* beds), and early Tiffanian stages of the Western United States and Canada (New Mexico, Wyoming, Montana, North Dakota, Alberta).

Revised diagnosis: [14] Small to middle-sized pantodonts, ambulatory, having full placental dentition.

DENTITION

Lower incisors conical, sometimes with lateral crest on each side; increasing in size from I_1 to I_3. Lower canines moderately enlarged, sub-circular in cross section and in *Pantolambda* with tip pointing slightly backward; canine lacking posterior blade and anterointernal wing. P_1 one-rooted and tending to be elongated in an anteroposterior direction. Diastema, if present, posterior to the P_1. *Crista obliqua* of P_{2-4} usually reaching the lingual margin of the tooth. Molar and pre-molar cusps dilambdodont; paraconids of lower premolars and molars approximately equivalent to metaconids in

---

[14] Based largely on *Pantolambda bathmodon*, AMNH 16663; *P. cavirictus*, AMNH 963, and 17034, USNM 21327; and *Caenolambda jepseni*, PU 14863—for earlier diagnosis of this family, see Matthew (1937: 162).

size; lower molar trigonids distinctly shorter in antero-posterior diameter, and higher than talonids. Talonid of M_3 elongate anteroposteriorly, with posterior margin broken in a rather sharp angle, instead of being gently rounded as in the barylambdids. Upper incisors increasing in size from I^1 to I^3. Upper canine large and rounded or oval with the greatest diameter in the anteroposterior direction. P^1 one-rooted and set somewhat internal to posterior margin of canine. P^{2-4} three-rooted with single para-metacone V, protocone progressively increasing in size from front to back. M^{1-3} with parastyle extended labially beyond metastyle, so that the margin from the parastyle posterior along the external edge of the tooth forms an almost straight anteroposterior line. V's of paracone and metacone equally developed; protocone of M^{1-3} with small protoconule and metaconule, and large in proportion to paracone and metacone W. Large protocone of M^{1-3} gives molars a more quadrate or rectangular outline than in the molars of any other family of pantodonts; protocone both longer anteroposteriorly and wider transversely in proportion to metacone and paracone. Basal cingula of molar protocone in postitions of protostyle and hypocone well developed in this family, and in some species forming a pronounced cingular shelf encircling the lingual margin of the protocone. M^3 metacone reduced but showing traces of metastyle in some individuals, presumably a primitive condition.

JAW AND SKULL

Posterior extension of jaw angle reduced (except in *Pantolambda bathmodon*), coronoid process relatively short, mandibular symphysis fuses before eruption of permanent dentition. Symphyseal region of mandible relatively heavy for a pantodont, with moderate to extreme development of a horizontally extended flange on the anteroexternal margin of the jaw ramus. Premaxilla large, with facial extension; nares terminal; nasals elongate, expanded posteriorly and, except in *Pantolambda bathmodon*, narrow. Skull long and narrow posterior to the orbits, brain case small, with sagittal and lambdoid crests prominent, but not usually extending posterior to occipital condyles.

POST-CRANIAL SKELETON

Cervical centra not so long relative to size of vertebra as in some condylarths, such as *Ectoconus* and *Phenacodus*, but proportionally longer than in the Barylambdidae, corresponding in this respect to the Coryphodontidae and Titanoideidae. Centrum of axis comparatively shorter than in *Titanoides;* spine heavy and elongated. Seventh cervical not pierced by vertebrarterial canal. Thoracic vertebrae numbering not less than fourteen and relatively shorter than the five lumbars which increase in size posteriorly; three fused sacrals fit into pelvis by means of a joint between ilium and fused transverse processes of the anterior two sacrals. The limbs of the known pantolambdids lack many

of the distinctive specializations of the corresponding parts of the skeletons of the other families of pantodonts, but possess the following primitive mammalian characters adequately summarized by Osborn (1898b: 185) for *Pantolambda:*

Fore-limb strongly bent outwards at elbow (as in Creodonta and Carnivora), manus everted. Humerus with powerful deltoid, pronator (entepicondylar) and supinator (ectepicondylar) crests; ulna with a convex posterior border; carpus with an os centrale, an extremely small magnum and short trapezoid, causing the metacarpal IV to be inserted proximally between the trapezoid and magnum. Hind-limb straight with three trochanters upon the femur. Tibia with rudimentary spine, a very long cnemial crest and femoral facets approximate. Tibia articulating with calcaneum. Probably an *os tibiale.* Mesocuneiform short (analogous to trapezoid in the carpus) so that metatarsal IV articulates between ento- and ectocuneiforms (analogous to metacarpal IV). Articulation between tibia and astragalus slanting obliquely inwards, very limited in extent, bounded posteriorly by astragalar foramen which issues posteriorly between the ectal and sustentacular facets (as in Creodonta). Astragalo-navicular head broad convex; two astragalo-calcaneal facets only. Fifth metatarsal curved with a prominent external tuberosity for the *peroneus brevis* muscle, as in the Bear.

As Osborn points out, all these characters are shared by *Periptychus.*

The major distinctive features of the pantolambdid skeleton are as follows: Scapula with distinct suprascapular notch; anterior and posterior blades wide, but not so wide as in Barylambdidae, blades constricted at base; acromion long; metacromion not pronounced, but forming broad posterior angle between spine and acromion; coracoid process large and curved slightly backward; vertebral border gently rounded, no sharp gleno-vertebral angle; entire bone about one-third longer than wide. Clavicle large and long, slightly curved; shaft round in cross section, not flattened; expanded at ends, broadest at sternal end. Manus of Pantolambdidae differing from that of *Periptychus* in having proximal phalanges shorter relative to metacarpals. Unguals elongate, somewhat flattened and fissured. Pes, with calcanium lacking internal tubercle.

PANTOLAMBDA Cope, 1882: 418

Type: *Pantolambda bathmodon* Cope.
Included species: *Pantalambda bathmodon, Pantolambda cavirictus, Pantolambda intermedius.*
Distribution: Torrejon stage; Upper Torrejon beds or *Pantolambda* zone, Nacimiento formation, San Juan Basin, New Mexico; Rock Bench level, Polecat Bench formation, Bighorn Basin, Wyoming; upper part of Lebo, Crazy Mountain Field, Montana (Gidley Quarry and other scattered localities) and in approximately the lower five hundred feet of the Melville formation, Montana.
Generic characters: The characters of this genus are extensively discussed by Matthew (1937: 163),

briefly they are: Comparatively small size, from that of a fox (*Pantolambda bathmodon*) to that of a tapir (*Pantolambda cavirictus*); incisors, above and below, small and simple; canines never flattened transversely —relatively larger than in barylambdids and smaller than in coryphodontids and titanoideids, lower canines distinctive, with backward tilted apex; diastema between P_1 and P_2, not between C and P_1—as in *Titanoides* and *Coryphodon;* molars and premolars lacking later specializations, M_3 with relatively large heel; muzzle comparatively short and small. Post-cranial skeleton generalized with little expansion of scapula and illium; limbs lacking specializations, humerus with entepicondylar foramen, tibia retroflexed, centrale of manus and astragalar foramen of pes present and relatively large.

Discussion: This genus includes the smallest, oldest and most primitive pantodonts. It approximates the structural ancestry of the order, but even within the genus, as would be expected of contemporary species, there are evident distinctions. *Pantolambda bathmodon* is much the most primitive of the three species. It shows resemblances (which are not present in any other pantodont) in the pes, limb bones, and pelvis in particular to protungulates such as the periptychids and phenacodonts. In fact, if the dentitions of *Pantolambda bathmodon* and *Pantolambda cavirictus* were not so similar, one might regard the considerable differences between the two species in post-cranial skeleton as warranting generic separation. Such a separation has been suggested by Patterson (1939a: 356) and may eventually be advisable but, pending the discovery of a really complete skeleton of *Pantolambda cavirictus* which would add an element of certainty to the many fragmentary specimens which have been referred to this species, an attempt to separate the two forms generically on the basis of skeletal characters seems unwise. *Pantolambda intermedius* is at present too poorly known to allow very informative comparisons between it and the other species of the genus. Its dentition, however, resembles that of *Pantolambda bathmodon* more closely than it does that of *Pantolambda cavirictus.*

Pantolambda bathmodon Cope, 1882

Figures 1, 2, 7, 10 and 16

Pantolambda bathmodon Cope, 1882, *Amer. Nat.* **16**: 418.

Type: AMNH No. 3956; part of a mandibular ramus including P_3-M_1.

Referred specimens: AMNH Nos. 2545, 2546, 2547, 2549, 2550, 2551, 2552, 3957, 3958, 16663, 16664, 16666; [14a] USNM Nos. 5903, 5904, 15408, 15409, also

[14a] A number of other very fragmentary AMNH specimens have been assigned to *P. bathmodon.* Some of these are indeterminate and the best of them shows nothing not better preserved in the referred specimens above. For the sake of completeness, however, these specimens should be listed. They are AMNH 2523, 2548, 2554, 4045, 15934, 16665.

31-49, 51-49, 160-36; KU Nos. 8067, 8068, 8069, 8070, 8071, 8072, 8073, 8074, 8075, 8076, 8077.

Horizon and locality: Torrejonian stage; upper Torrejon beds, Nacimiento formation, San Juan Basin, New Mexico. Probably occurs in the Medicine Rocks local fauna, Tongue River Paleocene, Carter Co., Montana, also Torrejonian stage.

Diagnosis: See Matthew (1937: 167).

Discussion: This species has been thoroughly treated by Matthew (1937), and since the time of his writing little material that amplifies his observations has been found. However, in spite of the long history of publication on this species, crown views of the complete upper and lower dentitions have never been figured. AMNH 16663 presumably preserves the entire upper and lower dentition, but the hardness of the matrix has to date prevented separation of the mandible. No one of the separated lower jaws in the Cope collection, at the American Museum of Natural History, contains a very complete dentition, but it is possible to reconstruct a composite picture of the teeth from the lower C to M_3. The best preserved upper dentition of this species, illustrated in figure 10, USNM 15408, includes the dental series from P^1 to M^3. The lateral view of the skull, included by Matthew (1937: fig. 39), of *Pantolambda bathmodon,* drawn principally from AMNH No. 16663, is somewhat misleading owing to the fact that the symphyseal region of the mandible and the anterior part of face have been crushed together. The illustration does not adequately correct for this distortion. The mandible in the symphyseal region and below the canine is usually much deeper than in this illustration and some individuals possess a strong anteroexternal flange, similar to that of *Pantolambda cavirictus.* For the same reason, the muzzle in Matthew's figure is not deep enough above the canine.

Pantolambda cavirictus Cope, 1883

Figures 1, 5, 7, 10 and 16

Pantolambda cavirictus Cope, 1883, *Proc. Acad. Nat. Sci. Phila.* **35**: 169; 1884, *Amer. Nat.* **18**: 1114, fig. 6.
Pantolambda bathmodon Osborn and Earle, 1895, *Bull. Amer. Mus. Nat. Hist.* **7**: 43-46, figs. 14-16.

Type: AMNH No. 3961, lower jaw with dentition lacking only crowns of incisors, left and right P_1, left condyle and parts of both angles.

Referred specimens: AMNH Nos. 962, 963, 964, 965, 2455, 2555, 2556, 3961, 3963, 3963b, 3964, 3967, 3986, 16042, 16723, 17034, 35727; USNM Nos. 15406, 15407, 6155, 9858 (single teeth), 21327; PU Nos. 13759, 16456.

Horizon and locality: Torrejon stage; Upper or *Pantolambda* zone, Torrejon beds, Nacimiento formation, San Juan Basin, New Mexico; upper part of Lebo formation, Crazy Mountain Field, Sweetgrass County, Montana (level of Gidley Quarry and above in approx-

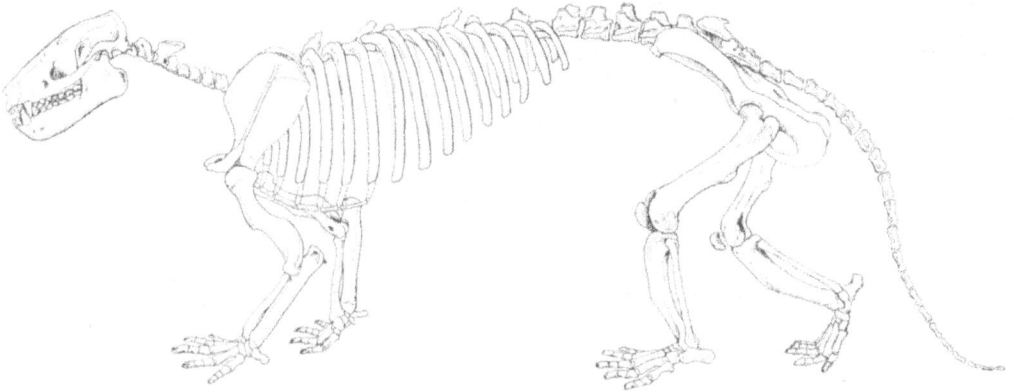

FIG. 2.   × 1/6.   *Pantolambda bathmodon* Cope, after Osborn 1898, with corrections from AMNH 16663.

imately the lower five hundred feet of the Melville formation) ; Fort Union formation, Billings Co., North Dakota ; Tongue River formation, Medicine Rock local fauna, Carter Co., Montana.

Specific diagnosis: Many observations on the characters of this species have been previously published, and since these are extensively reproduced in Matthew (1937 : 180–181), along with his own remarks on this form, they will not be quoted here. The new specimens recovered since 1937 and the necessity for correcting some conflicting statements in earlier literature on this species make a revised diagnosis and discussion advisable. For the sake of completeness, observations derived from earlier authors have been included in this diagnosis.

Revised diagnosis: Teeth and skull about one and one-half times as large, and vertebrae, limb bones, and sacrum about twice as large as corresponding elements of *Pantolambda bathmodon*. Incisors above and below agreeing with those of *Pantolambda bathmodon* in their relative sizes, which increase from first to third, differing above in being relatively larger compared to the other teeth and more spatulate. Upper canine relatively larger than in *Pantolambda bathmodon* and oval (instead of subcircular) at base. Lower canine subcircular at base and broad enough to expand the alveolar border which is much narrower posterior to the canine; as in *Pantolambda bathmodon*, apex directed slightly posteriorly. P_1 one-rooted and followed by a long diastema; P_2 slightly smaller relative to P_3 than in *Pantolambda bathmodon*. Details of lower premolar and molar series closely corresponding except in size to *Pantolambda bathmodon*, although talonid of M_3 of *Pantolambda cavirictus* somewhat smaller relative to trigonid than in the former species. Upper P^1 relatively as large as corresponding tooth in *Panto-*

lambda bathmodon,[15] but situated somewhat more internal to posterior margin of upper canine. P^2 sometimes a little larger in proportion to P^3 than in *Pantolambda bathmodon*, usually agreeing with the latter species in possessing a very small protocone on the P^2 which gives it a comparatively shorter transverse diameter than in the succeeding teeth; anterior and posterior cingula of protocone of upper premolars and molars increasing in heaviness from front to back so that M^{1-3} may have a cingular shelf completely encircling the protocone; agrees with *Pantolambda bathmodon* and other pantolambdids in having reduced molar metastyle not projecting beyond the labial margin of the tooth, posterior to the parastyle.

SKULL

Mandible showing development of anteroexternal flanges on horizontal rami[16] intermediate in heaviness between *Pantolambda bathmodon* and *Caenolambda jepseni*. Coronoid process relatively shorter than in *Pantolambda bathmodon* and angle of jaw less prominent; symphysis heavier and depth of ramus relatively greater than in the smaller species. Cranium more elongate than in *Pantolambda bathmodon*. Premaxilla massive, upper incisors relatively large, nasals long and narrow, maxilla high anteriorly supporting huge canine root, zygomatic arch rising above the P^4 somewhat more flared and slender than in *Pantolambda bathmodon*; frontal, sagittal, and occipital crests and other proportions and features of the skull much as in *Pantolambda bathmodon*.

---

[15] Matthew's observation (1937 : 182) that the P^1 of *P. cavirictus* is proportionately smaller than that of *P. bathmodon* is based on a much worn and broken specimen, AMNH 2555; see instead, USNM 21327 in fig. 10.

[16] Referred to by Cope (1883 : 169) as the "lateral prominence of the inferior edge of the ramus. . . ."

POST-CRANIAL SKELETON

Vertebrae closely resembling those of *Pantolambda bathmodon* and *Caenolambda jepseni*, and intermediate between them in size; spine of axis large and elongate; tail long and heavy, and judging from the rate of decrease in size of the caudal centra in partial specimens, with about fifteen vertebrae. Scapula, humerus, radius, and ulna differing in known parts only in size from those of *Pantolambda bathmodon;* entepicondylar foramen present. Manus only partially known; not distinct except in size, from *Pantolambda bathmodon.* Pelvis less acuminate than that of *Pantolambda bathmodon* and much less like that of the phenacodonts; with some expansion at the anterior end of ilium, but not so broad and flattened as in larger pantodonts, *Titanoides, Barylambda* and *Coryphodon.* As in *Pantolambda bathmodon* an oblique ridge can be traced on ventral surface of ilium, but in *Pantolambda cavirictus* it is less well developed. Pes more robust than that of *Pantolambda bathmodon* with relatively thicker metatarsals; some ungual phalanges barely fissured; much longer than wide.

Discussion: This species is known from a relatively large number of specimens, none of which is very complete. Consequently, discussion of proportions and variability remains extremely conjectural. It seems probable that if this form were better known it would be possible to distinguish more than one species, judging by the differences in size of various specimens and in certain characters of the dentition, such as development of the cingula of the upper molars, size and proportion of the upper canine, and size and position of the P_1. Distinctions of this same degree in the dentitions of other pantodonts have been associated with striking differences in the rest of the skeleton such as those between *Barylambda faberi* and *Leptolambda schmidti.*

Since little is known of the size range in *Pantolambda cavirictus,* it seems unwise to subdivide the species on the basis of size alone, but it should be pointed out that the type lower dentition, *Pantolambda cavirictus,* AMNH 3961, measures 140 mm. from the anterior margin of the canine to the posterior margin of the M_3, while the corresponding measurement on the upper dentition of AMNH 963, is much less, being 106.2 mm. In other pantodonts, for instance *Titanoides primaevus,* CNHM 15520, corresponding measurements along the tooth row, from the anterior margin of canine to posterior margin of M3, above and below are almost exactly equal. If this proportion is also true for *Pantolambda cavirictus* (and although there are no associated upper and lower dentitions complete enough to prove that it is, this seems a safe assumption) the above measurements indicate either that there was a size range in the species as great as twenty-five per cent, or that these specimens represent two separate species. At present it seems best to consider this size difference within the species range, for even greater ranges are known in many living mammals such as the jaguar and puma and are suspected for *Barylambda faberi.* Matthew (1937: 182) also noticed the disproportion in size between the most complete maxillae and mandible of *Pantolambda cavirictus* remarking that:

The skull, as figured by Osborn, has been restored from several separate pieces and is too narrow in the palatal region and muzzle. The actual width, judging from the width of the symphysis in the type lower jaw, should give almost twice the width between the canines shown in the skull as figured. . . .

Matthew's observation that the muzzle of AMNH 963 is too narrow as restored is certainly true, but it is also evident that the symphyseal region of the type lower jaw was mistakenly restored with too great a width. Even after both errors are corrected, the muzzle is too narrow for the jaw, further confirming the marked difference in size between the type mandible and the maxillae of AMNH 963. The only evidence that these two specimens do belong to the same species is that a lower premolar, very similar to those of the type, was found with AMNH 963. Several partial upper dentitions have since been found in the upper beds of the Torrejon formation of the San Juan Basin of New Mexico, but none of these was associated with lower teeth. One of these dentitions, AMNH 2555, from Arroyo Torrejon, is slightly larger than AMNH 963 and differs from it in having continuous cingula encircling the bases of the protocones apparently on all teeth from P^3 to M^3. A maxilla with $C–M^3$, USNM 21237 is illustrated here in figure 10. Two other isolated teeth from Arroyo Torrejon USNM Nos. 15406 and 15407 a lower P_4 and lower M_2 respectively, are at opposite extremes of size for this species, the P_4 being larger than in the type lower jaw, while the M_2 is closer to the size expected for AMNH 963.

An upper $M^{2–3}$ from the Gidley Quarry, AMNH 35727, is like small *Pantolambda cavirictus* specimens, but in the large size of protocone and straight labial margin of M^2 posterior to the parastyle resembles *Caenolambda jepseni* (described below). Two teeth possibly associated with this specimen are quite problematic; one may be a lower canine, the other an I^3, but if so, the I^3 is extremely large for *Pantolambda cavirictus.* This specimen, as well as USNM 6155 and 9858, single upper molars, mentioned by Simpson (1937a: 271), all fall within the range of *Pantolambda cavirictus,* as here defined.

Wortman (1897: 82) referred a pelvis with associated thoracic, lumbar, fragmentary sacral and caudal vertebrae to *Psittacotherium,* but qualified his reference as follows:

Among the specimens collected last summer by the Expedition, is one found by my assistant, Mr. Barnum Brown, in the Torrejon Beds or the Upper Puerco, consisting of two posterior dorsal, three lumbar and nine caudal vertebrae, together with a nearly complete pelvis. It was thought at first that these bones belonged to a large Creodont, probably a species of *Dissacus,* but they were so much broken that it was impossible to form any correct judgment of their true characters at the time of collection. Since they have been cleaned and mended, it is now very evident that they do not belong to any known species of Creodont. The only other species occurring in this horizon to which they could be referred on account of size are *Pantolambda cavirictus* and *Psittacotherium multifragum.* A comparison with the corresponding bones of the smaller species of *Pantolambda* shows such great and fundamental differences that it is certain that the specimen in question does not belong to *Pantolambda.* There remains then only *Psittacotherium multifragum* to which the specimen can be referred, and if we can judge from the peculiar characters which these bones present, I think there can be no mistake in referring them to this species.

Later students, however, have not seen any compelling reason for believing that this specimen belongs to a taeniodont. Matthew (1937: 266) stated his position clearly, as follows:

Wortman has referred to *Psittacotherium* a pelvis with posterior dorsal, lumbar, and anterior caudal vertebrae associated. We have not succeeded in finding any of the fundamental differences which he states separate it from *Pantolambda.* The characters of the pelvis are not unlike those of *Pantolambda bathmodon;* the form of the ilium, the elongate ischium, separate and long pubis, are present in *Pantolambda,* and as far as appears, it may belong to *P. cavirictus.*

Matthew further stressed in no uncertain terms his doubt that this pelvis has any particularly edentate (that is, taeniodont) characters or any features which would weigh in favor of the specimen being *Psittacotherium multifragum* rather than *Pantolambda cavirictus.* He concluded that the specimen was indeterminate.

Subsequent to Matthew's writing, pelves and vertebrae of two other Paleocene pantodonts, *Titanoides primaevus* and *Barylambda faberi,* have been discovered; and these specimens confirm in large part by their similarity Matthew's suggestion that the pelvis described by Wortman might be of *Pantolambda cavirictus.* However, evidence sounder than this has been supplied by the discovery of a new pantolambdid species *Caenolambda jepseni* with associated thoracic, lumbar, and caudal vertebrae which are strikingly like those associated with the pelvis, AMNH 2455, described by Wortman. The major difference in the two sets of vertebrae is that of size only.

The importance of this pelvis and associated vertebrae for a better understanding of the taxonomic relationships of the pantodonts is considerable, and although this specimen cannot be referred with absolute certainty (in the absence of an associated dentition) to *Pantolambda cavirictus,* it seems most probable that the specimen does belong to this species, or to a closely related pantolambdid. This pelvis makes an ideal intermediate in a structural series running from *Pantolambda bathmodon* through *Titanoides* to *Coryphodon;*

although this, of course, does not imply a direct ancestor-descendent sequence in these forms, suggesting only a trend the pantodonts were undergoing, see fig. 16.

The pes of *Pantolambda cavirictus* has been considered in detail by Cope, Osborn, and Matthew; but it should be pointed out here that less is known of it than one would assume from the published photographs and drawings, for in the right pes, AMNH 3963, originally described by Cope, several of the mesial phalanges are restored in plaster, as are *all* of the unguals except the central one in the mount. This one ungual is barely fissured, but this slight fissuring cannot be considered a significant difference from *Pantolambda bathmodon,* since it may not be the central ungual, which in *Pantolambda bathmodon* and the barylambdids is the most deeply fissured phalanx. Associated with this specimen are elements of a pes, AMNH 3963*b,* of slightly larger dimensions than the foregoing tarsus, AMNH 3963; these larger tarsals and metatarsals show that some of the Torrejon pantolambdids must have been almost as large as *Caenolambda.* The metatarsals of AMNH 3963*b* are fully as long as those of *Caenolambda jepseni,* but not so robust, or wide.

Of several hundred identifiable mammalian specimens from the Rock Bench Quarry of the Bighorn Basin of Wyoming, only four can be attributed to pantodonts. These specimens are all referable to the Pantolambdidae and probably to *Pantolambda cavirictus.* They consist of two phalanges, part of a lower molar and a navicular. The phalanges are of *Pantolambda cavirictus* type but somewhat larger. The tooth fragment, PU 16456, is not really complete enough for a definite reference, but it agrees well in size and in the proportions preserved, with this species. The navicular is of less certain assignment and, while not noticeably distinct from a pantolambdid navicular in shape, it is larger than that of *Titanoides* or *Caenolambda,* which suggests that there were much larger pantodont species existing in Torrejonian times not yet represented by adequate remains.

Pantolambda intermedius Simpson, 1935

Pantolambda intermedius, Simpson, 1935, *Proc. U. S. Natl. Mus.* 83: 244.

Type: USNM No. 8384. Left lower jaw with M_{1-2} alveoli of C-P$_4$, associated with symphysis fragment with right I_{1-2} and alveoli of left I_{1-3}.

Referred specimens: AMNH Nos. 35722, 35723, 35724, single premolars. USNM No. 9598.

Horizon and locality: Torrejonian stage, Lebo formation, Gidley Quarry, NW 1/4, NE 1/4, Sec. 25, T 5 N, R 15 E., Sweetgrass County, Montana.

Author's diagnosis: Intermediate in size between *P. bathmodon* and *P. cavirictus.* P$_1$ with one large root, close to canine, followed by short diastema. P$_{2-4}$ two-rooted. Lower molars closely resembling those of *P. cavirictus* but entoconid more distinct.

Discussion: An examination of the specimens of *Pantolambda bathmodon* in the Cope collection at the American Museum of Natural History indicates that a short diastema frequently occurs behind the P_1 in this species. The diastema, and the width of the alveolar border are proportionately the same as the diastema and alveolar border of *Pantolambda intermedius* behind the P_1. The corresponding diastema of *Pantolambda cavirictus*, on the contrary, differs from that of the two smaller species of *Pantolambda* in its relatively greater width and pronounced constriction of the alveolar border posterior to the P_1. The incisors of the type lower jaw of *Pantolambda intermedius*, although only partially erupted, are conical, as in *Pantolambda bathmodon*, instead of being somewhat spatulate, as in the larger pantolambdids, including *Pantolambda cavirictus*. Three isolated lower premolars of *Pantolambda* type which are also from the Gidley Quarry, AMNH Nos. 35722, 35723, and 35724, are all of appropriate size to pertain to this species, and, like the premolar, USNM No. 9598, described by Simpson (1937a: 270) as referable to *Pantolambda intermedius*, they are narrow and the metaconid crest is quite small or absent. Another lower premolar from the same quarry is much more quadrate at the base, with a well-developed metaconid and heel. In all respects it is very close to the P_4 of the Titanoideidae, and is unlike the premolars mentioned above which presumably belong to *Pantolambda intermedius*. Because of these titanoideid features, it is more probable that this premolar belongs to *Titanoides simpsoni* (described below) which has upper molars that agree well with it in size.

Altogether in the characters of the diastema, incisors, and lower molars of the type specimen, as well as in the positioning of the entoconid of a referred lower premolar USNM 9598, see Simpson (1937a: 270), *Pantolambda intermedius* is much closer to *Pantolambda bathmodon* than to *Pantolambda cavirictus*. If it were not for the difference in absolute size, the two smaller species of *Pantolambda* would almost certainly be considered within the range of variability of one species. *Pantolambda cavirictus* and *Pantolambda bathmodon* are distinctly different in post-cranial skeleton, a fact which gives strong support for their generic separation, but in characters of the dentition the two species are less clearly differentiated. Nevertheless, in those features of the lower dentition in particular in which the two species, *Pantolambda bathmodon* and *Pantolambda cavirictus* differ, *Pantolambda intermedius* resembles the former species. Since the post-cranial skeleton of *Pantolambda intermedius* is unknown with any certainty, [see Simpson (1937a: 271)], the most definitive comparisons cannot be made.

Pantolambda bathmodon is not known from the Torrejonian equivalents of the Bighorn Basin of Wyoming (Rock Bench Quarry), or from the Crazy Mountain Field of Montana. It is consequently possible to conjecture that the size difference between the two species, *Pantolambda bathmodon* and *Pantolambda intermedius*, may be due to the operation of the ecologic principle known as Bergmann's rule. This rule briefly stated is that the more northern species or subspecies of a given type of animal tend to be larger than their more southern relatives. Other possible interpretations of the size differences are: (1) that the two species are not contemporary, and that *Pantolambda intermedius* is a larger species descended from *Pantolambda bathmodon;* or (2) that the two forms are synchronous but with specializations separating them ecologically which are not evident in the known parts. It is also just possible that the lower jaw, AMNH 962, from the Cope collection of the San Juan Basin of New Mexico, considered by Matthew (1937) to be an immature dentition of *Pantolambda cavirictus*, belongs to *Pantolambda intermedius*. The bone of the horizontal ramus of this specimen is poorly preserved and only two teeth remain. It is not clear, therefore, whether the M_3 alveoli are present, though x-rays suggest that they may be. If so, this specimen from the San Juan Torrejonian could be referable to *Pantolambda intermedius* and would show that this species could be contemporaneous with the other two species of *Pantolambda* in New Mexico.

CAENOLAMBDA GAZIN, 1956: 48

Type: *Caenolambda pattersoni* Gazin, 1956.

Included species: *Caenolambda pattersoni, Caenolambda jepseni.*

Distribution: Tiffanian stage; lowermost Silver Coulee beds, Polecat Bench formation, Big Horn County, Wyoming; Saddle locality, Bison Basin, Fremont County, Wyoming.

Author's description: Skull with elongate cranium, strong, arched sagittal crest, broad frontals, narrow nasals and heavy canines resembling the *Titanoides* group. Upper cheek teeth, though comparatively small, are anteroposteriorly shortened and transversely broad as in the *Barylambda-Haplolambda* group, but with molars M^1 to M^3 about equalling one another in size.

Generic characters: To Gazin's original description, based on the type skull, may be added the following dental and osteological characters derived from the partial skeleton, in the Princeton collection, of the type of *Caenolambda jepseni.*

DENTITION

Unreduced. Incisors relatively small, pointed, more spatulate than in *Pantolambda bathmodon*, increasing in size from front to back. Canines much enlarged, sub-circular in cross section below, but above oval at the base, elongate and tusklike. First premolar below, one-rooted and elongate anteroposteriorly; not followed by a diastema as in *Pantolambda;* P^1 one rooted, lacking protocone, moderately compressed transversely. P^{2-4} above and below increasing in size from front to back, but more nearly equal to each other in size than in *Pantolambda*. Molars above with paracone and metacone W and protocone V; lower molars with trigonid and talonid V; trigonids below more compressed anteroposteriorly and higher than talonids as in *Pantolambda*. Protocone of upper molars larger relative to the paracone and metacone W than in *Pantolambda*, development of protocone cingula above the same as in the latter genus. Size: corresponding measurements of the teeth about two-and-one-half times greater than in *Pantolambda bathmodon* and twenty to thirty per cent greater than in *Pantolambda cavirictus*. Relative sizes of molars to each other, above and below, as in *Pantolambda*.

SKULL

Mandible with massive symphysis and heavily developed anteroexternal flanges on horizontal rami (complete fusion of mandibular symphysis), coronoid process short, angle of jaw not greatly extended posteriorly. Cranium elongate with terminal nares and deep, massive muzzle supporting enlarged canines; nasals narrow; frontal, sagittal, and occipital crests high; brain case small; skull with pronounced postorbital construction; zygomatic arches moderately heavy, with anterior origin of arch above the P^2, instead of rising above the M^1 or P^4 as in *Pantolambda*.[17] Occipital condyles small relative to size of skull.

POST-CRANIAL SKELETON

Centra of cervicals relatively short, as in periptychids, but longer in proportion to the thoracic and lumbar vertebrae than in the Barylambdidae. Thoracic and lumbar vertebral centra about equal in length. More than fourteen thoracic, and at least five lumbar vertebrae; sacrals not known; anterior caudals large, presumably supporting a well-developed tail, as in *Pantolambda*. Scapula with wide anterior and posterior blades of about equal width, pronounced suprascapular notch, spine high and prominent, with a tuberosity, not known in other pantodonts, situated one third of the length down from the superior border to the acromion, superior border thickened; acromion long and moderately wide. Coracoid process well developed. Clavicle long, subcircular in cross section, or oval, but not flattened; somewhat curved, and wider at sternal end. Ribs, particularly the anterior ones, flattened at extremities, but subcircular near the proximal end and gently curved. Sternal elements, pelvis, and fore limb unknown. Hind limb: Femur stocky; corresponding, except in size, to that of *Pantolambda*, shaft relatively shorter (compared with the size of the head, great trochanter and distal condyles), than in *Coryphodon* or the barylambdids; third trochanter differing from *Titanoides* in its large size and position just above the middle of the shaft, connected to the great trochanter and external condyle by ridges as in *Barylambda*

faberi; second trochanter of average pantodont size. Tibia resembling that of *Pantolambda*, and differing from those of other pantodont genera in having a relatively high cnemial crest, internal malleolus large; fibula with ovate cross section, relatively short shaft and apparently with large articular ends.

Discussion: The species of *Caenolambda* have greater dental and osteological specializations than those of *Pantolambda*, and are also of larger size and, as far as is known, slightly later in time than any member of the latter genus.

Hence, one of the species of *Pantolambda* may be ancestral to *Caenolambda*. In many of the features in which *Pantolambda cavirictus* differs from *Pantolambda bathmodon*, it foreshadows characters of *Caenolambda*. Such linking features include: the anteroposterior elongation of the base of the upper canine, occurring in some specimens of *Pantolambda cavirictus;* the presence of a cingular shelf on the protocone of the upper molars of a few individuals of *Pantolambda cavirictus*, as in *Caenolambda pattersoni;* the general approach to quadrate form in the upper molars of both genera; the greater size of the upper molar protocone relative to paracone and metacone in both forms, compared to *Pantolambda bathmodon;* the similarities of the lumbar and caudal vertebrae—the principal difference being that of size alone. The tarsus is also similar in the two forms, but in *Caenolambda* the metatarsals are thicker and the astragalus is more advanced than in *Pantolambda cavirictus* in that it lacks as distinct a pit anterior to the trochlear surface. Consequently, the head is less well defined. Because of these similarities *Pantolambda cavirictus* is most probably closely related to, if not the direct ancestor of, *Caenolambda*. Since, however, both *Pantolambda cavirictus* and *Caenolambda* are based on partial skeletons, this postulated ancestor-descendent relationship cannot be firmly established. The manus is not completely known in either form, but judging from *Titanoides* some distinct differences in the manus could exist even when the dentitions of the two forms differ no more than they do in *Pantolambda* and *Caenolambda*. In the enlargement and anteroposterior elongation of the upper canine, *Caenolambda* resembles *Titanoides*, but this is the only prominent feature of the dentition which is more like *Titanoides* than *Pantolambda*. In all other major differences between the dentitions of *Pantolambda* and *Titanoides*, *Caenolambda* resembles the former genus to a much greater degree. Consequently, it does not seem probable that *Caenolambda* is a forerunner of *Titanoides*. That the similarities in the upper canines of the two genera are due to parallelism is further confirmed by the fact that the functioning of the lower P_1 in *Caenolambda jepseni*, which is elongated anteroposteriorly and shears against the back of the upper canine,[18] is quite different from that of *Titanoides*.

---

[17] The cavity anterior to the orbit, seen in Gazin's illustration of *Caenolambda pattersoni* (1956: pl. 13) is almost certainly an artifact.

[18] See figure 5.

Caenolambda pattersoni Gazin, 1956

Figure 1

Caenolambda pattersoni Gazin, 1956, *Smithsonian Misc. Coll.* 131 (6) : 48–50, pls. 12, 13, 14.

Type: USNM No. 21036, skull lacking zygomatic arches and mandible.

Horizon and locality: Lower Tiffanian stage, vicinity of Saddle locality, south rim of Bison Basin, Sec. 28, T 27 N, R 95 W, Fremont County, Wyoming.

Specific characters: Author's diagnosis: Length of skull greater than that of *Haplolambda quinni,* but less than *Titanoides primaevus.* Much smaller than *Barylambda faberi.* Cheek teeth comparatively small.

Revised diagnosis: [19] Upper canines large, tusklike, anteroposteriorly elongated at the base, and broader and longer at the base than in *Caenolambda jepseni.* $P^2 < P^3 < P^4$, but P^{2-3} relatively greater in transverse diameter in proportion to P^4 than in *Pantolambda.* Premolars narrower transversely and molars wider transversely than in *Caenolambda jepseni.* M^2 bearing a well-developed, cingular shelf on the lateral and lingual sides of the protocone—not present in *Caenolambda jepseni.*

Discussion: This recently discovered form from beds probably of early Tiffanian age in the Bison Basin of Fremont County, Wyoming, is a most interesting addition to the pantodont phylum; but in the case of this specimen, as is so often true of vertebrate fossils, the preserved parts raise as many questions as they answer. The type specimen, a single skull lacking zygomatic arches and mandible, is unlike any other pantodont of known Tiffanian age. *Caenolambda jepseni* would, of course, be an exception if it were indisputably of Silver Coulee age, but the fauna associated with it is as much like that of the Rock Bench level of the Bighorn Basin as it is like the Silver Coulee fauna.

In describing this new form, Gazin (1954: 48) remarks:

Caenolambda presents a rather unusual combination of characters and does not closely resemble any of the previously described genera. Nevertheless, in a general way the skull is apparently more like *Titanoides* than *Barylambda* or *Haplolambda.*

Gazin's further observations make it clear that the form is not to be allied with the Barylambdidae. He also noted several features of the dentition of *Caenolambda pattersoni* which resembled those of both *Titanoides* and *Pantolambda.*

It now seems certain that the new species *Caenolambda jepseni* from the Polecat Bench formation, Big Horn County, Wyoming, represents a related but distinct form, only slightly smaller than *Caenolambda*

pattersoni, and possessing an upper dentition in many ways intermediate between *Pantolambda cavirictus* and *Caenolambda pattersoni.*

These three species agree in possessing the following complex of characters: anteroposteriorly elongated upper canines, relatively shorter than in *Titanoides;* protocones of P^{2-4} centered on the transverse axis of the tooth, not shifted anteriorly as in *Titanoides* and *Barylambda;* M^{1-2} with parastyle projecting but metastyle reduced; [20] protocone of M^2 broader anteroposteriorly than in the Titanoideidae or Barylambdidae, giving this tooth an almost quadrate form.

The presence of an elongate cranium, broad frontals, and narrow nasals mentioned by Gazin as features of resemblance to *Titanoides* does not necessarily ally *Caenolambda* with the former genus; for these three features are also present in *Pantolambda cavirictus.* Further, as Gazin (1956: 49) pointed out,

The molars of *Caenolambda* would appear to be about equal to one another in size as in *Pantolambda,* not showing the marked increase from M^1 to M^3 seen in *Titanoides,* or the reduction of M^3 noted in *Barylambda* and *Haplolambda.*

One of the more useful taxonomic characters in the Pantodonta appears to be the relative sizes of the upper molars. Consequently, the agreement of *Caenolambda pattersoni* and *Pantolambda cavirictus* in upper molar proportions may be more than usually significant. The ledge-like cingulum, noted by Gazin, encircling the lingual and posterior margin of the M^2 of *Caenolambda pattersoni* is not a common feature of pantodont molars. In *Titanoides,* for example, the corresponding cingula, present on the anterior and posterior margins of the M^{1-2} protocones, may unite across the lingual face of the protocone; but this relatively faint joining of the cingula never resembles the strong cingular shelf present on the M^2 of *Caenolambda pattersoni.* The only other individual Paleocene pantodonts known to possess similar ledge-like cingula on the protocone of the M^2 are two specimens of *Pantolambda cavirictus,* AMNH 2555 and USNM 21327. The premaxilla of *Caenolambda pattersoni,* although broken and distorted, is clearly much larger in proportion to the skull as a whole than it is in *Titanoides.* The ascending ramus of the premaxilla, unlike those of *Titanoides, Barylambda,* and with less certainty *Haplolambda,* is extended on to the external face of the rostrum. In the latter genera this process is reduced and does not extend across the side of the face. In the large size of the premaxilla and the facial extension of the ascending process, *Caenolambda pattersoni* resembles *Pantolambda bathmodon, Coryphodon* and probably *Pantolambda cavirictus.* [21] As a consequence of these observed similarities between *Caenolambda pattersoni* and

---

[19] The discovery of a second species of *Caenolambda,* makes necessary the redefinition of the specific characters of *C. pattersoni.* Since partial upper dentitions are the only comparable parts in the two species, the diagnosis of *C. pattersoni* is restricted to the upper teeth.

[20] M^1 is not known in *Caenolambda jepseni.*

[21] Although no *P. cavirictus* specimen preserves an attached premaxilla, the large detached premaxilla associated with AMNH 16723 indicates that this species agrees with *P. bathmodon* in having a facial extension of the premaxillary process.

the other large pantolambdids, *Caenolambda jepseni* and *Pantolambda cavirictus*, it seems reasonably certain that they form a closely related group, which in turn is allied with *Pantolambda bathmodon* more nearly than to the other divisions of pantodonts.

As Gazin noted, the preservation of detail in the type skull of *Caenolambda pattersoni* is not particularly good; but in those respects in which the dentition can be compared with *Caenolambda jepseni*, the differences between the two forms make specific separation advisable; at least until a more complete series of this type of pantodont has been discovered.

In the frontal region of the skull, above the orbits of *Caenolambda pattersoni*, are swollen areas, which are not as strongly developed in other pantodonts. The rugosity of these enlarged areas seems to be natural, and in a general way they resemble the bases for attachment of horny outgrowths. A similar, but much less pronounced, temporal ridge is present in *Barylambda* as well. Possibly the development of these bosses proceeded with age, for the much worn dentition of the type of *Caenolambda pattersoni* may be taken to indicate advanced age.

Caenolambda jepseni sp. nov.[22]

Figures 1, 5, 6, 7, 9, 10, 13

Type: PU No. 14863: Lower jaw with complete dentition represented on one side or the other, upper right I, C, P^{1-4}, M^2, left P^3, M^2; lacking skull; twenty-six vertebrae, including seven cervical, twelve thoracic, five lumbar, and two caudal; both clavicles, right scapula, parts of about twelve ribs, left femur, part of right femur, left tibia and part of fibula; much of right pes.

Horizon and locality: Stratigraphic position and associated fauna suggest an age about at the boundary between Torrejonian and Tiffanian stages, but more probably to be associated with Torrejonian; NE 1/4, Sec. 18, T 55 N, R 96 W, apparently slightly higher than Rock Bench level, Polecat Bench formation, Big Horn County, Wyoming.

Specific diagnosis: Dental formula I ?3/3, C 1/1, P 4/4, M 3/3; larger than *Pantolambda cavirictus* in known parts, with dentition differing from this species as follows: *Caenolambda jepseni* lacking pronounced diastema between P_1 and P_2 and having instead an anteroposteriorly elongated P_1, which appears to have sheared against posterior face of upper canine; P_{2-3} larger relative to P_4; anteroexternal flange of jaw rami much more developed; upper P^2 larger relative to premolar series; upper canine much more elongated anteroposteriorly. I_2 slightly larger than I_3, incisors not increasing in size from I_1 to I_3 as in barylambdids and *Pantolambda bathmodon*. Upper molars not as broad transversely as in *Caenolambda pattersoni*, and lacking

[22] Named for Dr. Glenn L. Jepsen under whose direction this specimen and many others included in this paper were collected.

the well-developed cingular shelf of protocone of upper molars of latter species. Parastyle of M^2 much smaller relative to tooth size than in *Caenolambda pattersoni*.

Discussion: The distinctions of the dentition of *Caenolambda jepseni* from other pantolambdid species have been pointed out in the diagnosis of this species. Neither these dental characters nor the known postcranial skeleton suggests that *Caenolambda jepseni* possesses any characters which associate it with the non-pantolambdid pantodonts. Its most unusual dental feature, not known in any other pantodont, is the large anteroposteriorly elongated sectorial or shearing P_1. The superficial appearance and positioning of this tooth is not unlike that of some cercopithecoid monkeys, for example *Papio porcarius* and *Cynopithecus niger*, except that in these species the tooth is two-rooted and is, of course, the P_2. This convergence, to the extent that it is real, implies similar feeding habits.

The possible phylogenetic relationships of this genus with *Pantolambda cavirictus* have already been discussed and the suggestion made that the latter form foreshadows *Caenolambda* in many respects. In one feature, the lower dentition of *Caenolambda jepseni* is more like that of *Pantolambda bathmodon* than that of *Pantolambda cavirictus*; this is in the position of the P_1. The former species agree in having the apex of the P_1 shifted to the posterior end of the tooth; in the anteroposterior elongation of this tooth; and in lacking a broad diastema posterior to the P_1. This sectorial specialization in *Pantolambda bathmodon*, however, is not nearly so far advanced as it is in *Caenolambda jepseni*. Unfortunately, the type and most complete lower dentition of *Pantolambda cavirictus* lacks the P_1, but the close approximation of the P_1 alveolus to the posterior face of the lower canine, and the large diastema following the alveolus at least suggest that the apex of the P_1 in *Pantolambda cavirictus* might have been at the anterior end of the tooth so that the upper canine could follow down the posterior face of the canine and across the P_1 to rest above the diastema when the jaw was closed. If the tooth had the conformation of the P_1 in *Caenolambda jepseni*, it would clearly have interfered with the free movement of the canines. The wear surface formed by the P_1 on the posterior face of the upper dentition of *Pantolambda cavirictus*, USNM 21327, suggests, however, that the position of the P_1 in this species may not always have been as close to the lower canine as it is in the type specimen. In any event these distinctions both real and conjectural, in the region of the canine P_1 of *Pantolambda cavirictus* and *Caenolambda jepseni*, and the more general resemblance of *Pantolambda bathmodon* to *Caenolambda jepseni* in this respect, throw some doubt on the derivation of *Caenolambda* from *Pantolambda cavirictus*. It remains uncertain whether the general resemblance is due to such a derivation or to parallelism. Figure 5 is intended to show more clearly these distinctions in the occlusion of pantodont teeth in the region just anterior and posterior to the canines.

The association of the type of *Caenolambda jepseni* with several species of mammals not known to occur above the Rock Bench level in the Bighorn Basin of Wyoming makes reference of this form to the Torrejonian stage plausible in the present state of knowledge. The stratigraphic position of the locality of the type of *Caenolambda jepseni* in the Polecat Bench formation is in the region of the transition from the Rock Bench level to the lowermost Silver Coulee beds. There are also the proximal and distal parts of a tibia in the Paleocene collections from New Mexico in the AMNH, No. 16753 which may belong to this species. The proportions of this specimen collected in 1913 from the upper Torrejonian formation of the Escavada Wash agree with the tibia belonging to the type of *Caenolambda jepseni* very closely, but the New Mexico specimen is slightly larger. AMNH 16753 is almost certainly too large to be within the range of *Pantolambda cavirictus* judging from the much smaller tibial fragments associated with known individuals of this species, for instance AMNH 965. Although this one bone is very slim evidence for the presence of *Caenolambda* in the Torrejon formation of New Mexico, there is no other known Torrejonian mammal large enough to have a tibia of this size, and the collecting data appear to be accurate.

A basic outline of the skeletal distinctions of this form has already been included in the revised diagnosis of *Caenolambda*. Detailed comparisons of this unique skeleton with other pantodonts are included in Section IV, The Pantodont Skeleton.

Caenolambda sp. indet.

Pantolambda sp. Russell, 1948: 153.

NMC 8870, a partial upper left M², lacking most of metacone.

Horizon and locality: Presumably Torrejonian or Tiffanian stage Paleocene, Upper Saunders beds, Saunders series of Central Alberta; found three miles east of Saunders Creek station in an outcrop on the Nordegg Branch of the Canadian National Railways.

Description: Quoted from Russell (1948: 153).

. . . outline of crown more rectangular than triangular, with anterior and posterior margins nearly parallel; external margin short, with parastyle projecting strongly on the external side, and having a slight hook shape on the anterior side; paracone (and metacone) narrowly crescentic; protoconule minute but distinct; protocone large, almost conical, anterior and posterior wings joining cingulum; cingulum continuous except on innermost slope of protocone, and with a hypocone-like style at juncture with posterior limb of protocone.

Discussion: This tooth was provisionally referred to *Pantolambda* by Russell, 1948, and he was entirely correct in stressing the similarities of this Saunders

Creek tooth to the upper second molars of *Pantolambda*. Although this tooth, NMC 8870, is twice as large as the M² of *Pantolambda bathmodon* and a half larger than the corresponding molar of *Pantolambda cavirictus*, in conformation it approximates the latter forms much more closely than it does this tooth in any of the barylambdids, *Titanoides* or *Coryphodon*.

The recent discovery of a new early Tiffanian pantolambdid genus and species, *Caenolambda pattersoni* Gazin (1956: 48) greatly clarifies the problem of the taxonomic affinities of the Saunders Creek tooth. NMC 8870 almost certainly belongs to the genus *Caenolambda* for it agrees with the M² of the two species of *Caenolambda* in all the features of this tooth in which the latter genus differs from *Pantolambda*.

As Russell (1948: 153) pointed out, the Canadian specimen differs from the M² of both species of *Pantolambda* in the more conical protocone and from *Pantolambda cavirictus* only in the more quadrangular outline of the tooth. In these two respects as well as in size, this tooth is closest to the M² of *Caenolambda pattersoni*, but the incomplete union of the basal cingula across the lingual face of the protocone is a feature of resemblance to the M² of *Caenolambda jepseni*. However, in the Saunders Creek tooth, the labial extension of the parastyle is greater than in *Caenolambda jepseni* and more nearly approximates the M² in *Caenolambda pattersoni* in this regard. Since the specimen is so incomplete, an assignment of it to either of the two species of *Caenolambda* is unwarranted, but comparisons show that this tooth is definitely outside the range of variability of titanoideid or barylambdid upper molars. It consequently can be referred to *Caenolambda* with confidence.

The possible correlations which this specimen suggests are with faunas of the latest Torrejonian or earliest Tiffanian stages. *Caenolambda* is known only from beds which are close in age to the boundary between these middle and upper Paleocene levels.

FAMILY **BARYLAMBDIDAE** PATTERSON, 1939

Type: *Barylambda* Patterson, 1937, *Field Mus. Nat. Hist., Geol. Ser.* 6: 229–231.

Included genera: *Barylambda, Haplolambda, Leptolambda,* and *Ignatiolambda.*

Distribution: Tiffanian stage of Wyoming and Colorado, early Eocene or latest Paleocene of Southern Mongolia.

Diagnosis of family: [28] Medium to heavy, graviportal pantodonts with full eutherian dentition (I¹ sometimes lacking). Lower canine ranging from very small and incisiform to moderately large; usually bearing an anterointernal wing, but never having a posterior

---

[28] Patterson's original definition of the family has here been emended in order to include the genera *Leptolambda* and *Ignatiolambda*—quoted section is from Patterson (1939a: 362).

blade. P¹ one-rooted; P²⁻⁴ lacking basal cingula and much less quadrangular at the base than in the other pantodont families, particularly the Titanoideidae. Very slight diastema anterior and posterior to the lower canine and P₁. Metaconid of P₂₋₄ directed backward and paraconid wing anteriorly. Premolars below increasing in size posteriorly. Trigonids of lower molars greater in anteroposterior diameter than talonids, metastylid crests extending backward into talonids. Talonids of M₁₋₂ in *Barylambda* as wide or wider transversely than trigonids. M₁ relatively longer and wider than in pantolambdids; increase in size from M₁₋₃ not as great as in *Coryphodon*. Cuspules usually present in the molar talonids, particularly on the M₃. Upper incisors increasing in size markedly from I¹⁻³, I¹ sometimes absent; I³ (*Barylambda* only) large, moderately caniniform. Upper canine small to medium sized. P¹ one-rooted, lacking protocone, slightly elongated, much as in pantolambdids. P²⁻⁴ with parametacone V large and with transverse midline directed anteriorly; anterior and posterior crests of protocone present. Slight diastema present anterior and posterior to P¹. M¹⁻³ with paraconule and metaconule distinct, M¹ usually large relative to M²⁻³ but not in *Barylambda*, largest only in *Haplolambda*. The author's diagnosis continues:

> Skull with anterior nares terminal; muzzle not expanded; facial region short, wide, deep; nasals very wide; greatly expanded posteriorly; premaxillaries weak, not sutured medianly, ascending rami barely appearing on side of face and not reaching nasals; skull roof extremely wide and somewhat flattened across orbits; temporal ridges strong; zygomatic arches but little bowed outwardly; cranium low; occiput semi-circular, condyles extending well beyond it posteriorly. Symphysis of mandible long, sloping; coronoid process high; angle large.
>
> Cervicals with exceedingly weak neural arches; dorsals and lumbars very short in comparison with those of Coryphodontidae; sacrals with high spines, transverse processes and ribs greatly expanded and encroaching upon gluteal face of ilium; tail long, massive, provided with chevrons, deeper than wide anteriorly, spines of anterior caudals high, centra notably longer than those of dorsolumbars, neural arches complete on anterior half of series.
>
> Clavicle present, well developed. Scapula almost as wide as long; suprascapular border extensive, gently arched; acromion process long, robust. Head of humerus large, very deep antero-posteriorly, without anterior groove, articular surface not extending on to great trochanter; deltoid crest very strong, wide; entepicondylar foramen present; manus with large separate centrale; magnum articulating with Mc II; phalanges shortened; unguals broad, flat, those of digits II–IV deeply fissured.
>
> Pelvis exceedingly broad; ilium shorter antero-posteriorly than in *Sparactolambda* [*Titanoides*] or *Coryphodon*, antero-external angle considerably more anterior in position than genera just mentioned and bearing a plate-like process, ilium notched medially by gluteal process of sacrum. Femur broad, flattened antero-posteriorly, no pit for ligamentum teres in head. Tibia and fibula much more massive than in Coryphodontidae. Pes with heavy tarsus, relatively slender metatarsus and phalanges; astragalus without neck; calcaneum with very robust tuber calcis, fibular facet

small or absent; distal tarsals not displaced on metatarsals; phalanges shorter than in *Pantolambda*, longer than in *Coryphodon*; unguals small.—Patterson (1939a: 362).

Discussion: The large number of reasonably complete lower dentitions of *Barylambda faberi* and *Leptolambda schmidti* now known make it possible to estimate the approximate range of variability of proportion in the teeth of these two species. The logarithm of ratio diagram (fig. 9) in section A of the Appendix, shows that as in *Coryphodon*, the most variable tooth among the Barylambdidae is the canine. Moreover, as is clearly demonstrated in this diagram, the over-all proportions of the teeth in the barylambdid genera are remarkably similar, much more so than was expected when the original calculations were done, considering that breakage, distortion, erosion, sampling errors, and individual variation are all acting to distort the original ratios of tooth proportion among the specimens compared.

Figure 9 also indicates that the four barylambdid genera agree roughly in the way in which they all differ from the largest known pantolambdid *Caenolambda*, used as the standard in this diagram. This agreement in linear measurements further confirms the observed structural similarities which were the original basis for placing these genera in a distinct family. Furthermore, this diagram tends to confirm the idea that the two smaller barylambdid genera, *Ignatiolambda* and *Haplolambda*, although close to *Caenolambda* and *Pantolambda cavirictus* in size, do not link the barylambdids with the pantolambdids. As far as tooth proportion is concerned, they are as typical of the barylambdids as are either of the larger members of this family.

The agreement in linear proportions of the teeth, as well as the many observed similarities in the dentition and skeleton of the barylambdids, shows that the four genera belonging to this family are rather narrowly defined. Future discoveries might consequently make it possible to unite some of them, but at present they appear to be distinct and valid genera. It should also be emphasized that some of the species included in the Barylambdidae encompass considerable variability and they may well contain more than one species. The means of proving this supposition are not yet at hand, because of the fragmentary nature of some of the specimens which appear to be most divergent from species established here.

The recently described Asiatic pantodont *Haplolambda* [*Archaeolambda*] *planicanina* Flerow from beds of latest Paleocene or early Eocene age in the Nemegetu Basin of the southern Gobi, demonstrates that the pantolambdoids as well as the coryphodontoids occurred in both Asia and North America. Flerow, in 1952, established a separate family, Archaeolambdidae, for this species, but such a separation from the barylambdids is not warranted.

Haplolambda planicanina has no resemblance to *Pantolambdodon* that it does not share with the other pantodonts. As in the Pantodonta, but not *Pantolambdodon*, this species possesses a posterior premolar crest running from the protoconid to the metaconid and joined about the middle at a distinct angle by the *crista obliqua*. *Pantolambdodon*, however, has the dominant posterior premolar crest running from the protoconid directly backward to the posterior tip of the tooth, and this crest is joined at right angles by the metaconid crest. Consequently, *Haplolambda planicanina* provides no support for the idea that *Pantolambdodon* is a pantodont.

BARYLAMBDA Patterson, 1937

Titanoides Patterson, 1933: 415.
Titanoides Patterson, 1934: 71.
Titanoides Patterson, 1935: 143.
Barylambda Patterson, 1937: 229.

Type species: *Barylambda faberi* Patterson, 1933.
Included species: *Barylambda faberi.*[23a]

Barylambda faberi Patterson, 1933

Figures 1, 4, 5, 7, 9, 11

Titanoides faberi Patterson 1933, *Amer. Jour. Sci.* 5 (25): 415–425; 1934, *Proc. Amer. Philos. Soc.* 73: 71–101; 1935, *ibid.* 75: 143–162.
Barylambda faberi Patterson 1937, *Field Mus. Nat. Hist., Geol. Ser.* 6: 229–231.

Type: CNHM No. P-14637. Partial dentition and skeleton of immature individual.
Referred specimens: CNHM Nos. P-14898, P-14902, P-14944, P-14945, P-14946 (AMNH 32511), P-15075, P-25617, P-26110, P-26111.
Horizon and locality: Tiffanian stage; DeBeque formation, Plateau Valley fauna, Mesa County, Colorado.
Discussion: This species has been extensively described by Patterson (1939a) and in the earlier papers there cited; consequently, it will not be separately discussed here. Comparisons with this largest Paleocene pantodont have been made throughout the course of the present work and their results are included here principally in several of the figures, the section on pantodont anatomy, and the ratio diagrams, in Part A of the Appendix. It is curious that no specimens referable to this species have ever been found outside the Plateau Valley beds.

HAPLOLAMBDA Patterson, 1939

Haplolambda Patterson, 1939: 365.
Archaeolambda Flerow, 1952: 44.

Type species: *Haplolambda quinni* Patterson, 1939.
Included species: *Haplolambda quinni, Haplolambda planicanina.*
Distribution: North America: Tiffanian Stage: Silver Coulee beds, Polecat Bench formation, Big Horn

[23a] Specimens, horizon and locality are as for included species.

County, Wyoming. Plateau Valley local fauna, DeBeque formation, Mesa County, Colorado.
Asia: Earliest Eocene or latest Paleocene, lower Naran-Boulac horizon (post-Gashato), Nemegetu Basin, Southern Gobi Desert, Mongolia.
Diagnosis: I 3/3, C 1/1, P 4/4, M 3/3, no diastema. Smallest barylambdids with size range about as in *Pantolambda*. Canines not enlarged; lower molar talonids distinctly narrower and slightly lower than in other pantodonts; M^1 much longer and wider than P^4 compared to their relative sizes in other barylambdids; M^{1-3} series typically decreasing in size posteriorly, unlike other barylambdids where M^2 is largest; no constriction in depth of horizontal ramus of mandible beneath M_3.
Discussion: It is of some interest for the zoogeography of the Pantodonta that the recently discovered species *"Archaeolambda" planicanina* from the Nemegetu Basin of the Southern Gobi Desert is assignable to the genus *Haplolambda*. The lower teeth of the Asian species do not exhibit any definitely distinguishing features from those of *Haplolambda quinni*, except that they are in most measurements about a quarter smaller. The interruption in the posterior protocone crest of the upper P^3 and P^4 of *Haplolambda planicanina* is a real distinction from the condition here in the one upper dentition of *Haplolambda quinni*, but this feature is probably of little taxonomic importance as it also occurs sporadically in individual specimens belonging to *Leptolambda* among the barylambdids, and in *Titanoides, Eudinoceras,* and vary rarely in *Coryphodon*. If it were not for the great separation in space and also, apparently, in time, one might be tempted to question even a specific distinction, based on the existing material, between the Asian and North American species of *Haplolambda*. The two respects in which the jaw and dentition of *Haplolambda planicanina* differ most from that of *Haplolambda quinni* are that the horizontal ramus of the former species is comparatively more massive and that its absolute size is about twenty-five per cent smaller. Variations in absolute size among pantodont remains otherwise assignable to one species commonly approach twenty-five per cent, and differences in massiveness of the jaw ramus are also frequent in members of one pantodont species, so that neither of these characters may be very significant in establishing a distinction between these species of *Haplolambda*.

Haplolambda quinni Patterson, 1939

Figures 1, 5, 9, 11

Haplolambda quinni Patterson, 1939, *Field Mus. Nat. Hist., Geol. Ser.* 6: 365–368, figs. 106, 107, 108.

Type: CNHM No. P-15542. Anterior half of the skeleton.
Referred specimens: PU Nos. 16445, 16480, 16481.

Horizon and locality: Tiffanian stage; DeBeque formation, 15 feet above the base; Plateau Valley fauna, $1\frac{1}{2}$ miles north of Mesa, Mesa County, Colorado; Silver Coulee beds, Polecat Bench formation, SW 1/4, Sec. 22, T 54 N, R 95 W, and Divide Quarry NE 1/4, SW 1/4, Sec. 16, T 54 N, R 95 W, Big Horn County, Wyoming.

Author's diagnosis: Generally similar to *Barylambda* in the known parts, differing as follows: canines very small; M 3/3 reduced, metacone of M^3 vestigial, talonids of M_{1-2} narrower. Skull with much weaker temporal ridges. Mandibular ramus more slender, symphysis shorter. Shaft of ramus more slender, proximal and distal ends notably wider. Trapezium and Mc. I completely fused, the compound bone having a double articulation with Mc. II, the additional articular surface situated below, slightly behind and almost at a right angle to the usual one; Mc. V wider, nearly twice as wide as the other metacarpals, distal articular surface with median constriction.

Discussion: This species has previously been known from a single individual CNHM P-15542; consequently, it was particularly interesting to discover that a pair of lower jaws referable to this species was among the fragmentary specimens collected by the Princeton University field party of 1949 in the Silver Coulee beds of Big Horn County, Wyoming. The Princeton specimen, No. 16445, is slightly larger than the type of *Haplolambda quinni* but agrees well with it in the shallowness of the mandibular ramus, and in the general proportions of the teeth except that the lower molar talonids are somewhat broader than in the Chicago Museum specimen. Both of these lower dentitions are badly damaged, but the most reliable measurements have been plotted in the logarithm of ratio diagram (fig. 9). The two specimens of *Haplolambda* are close to *Ignatiolambda* in size but the lower P_{1-2}'s of the latter species are proportionally smaller than in *Haplolambda* and the M_{2-3} comparatively larger. In this respect the proportions of the lower teeth of *Haplolambda quinni* are less like those of the pantolambdids than they are in *Ignatiolambda*.

A left lower canine and part of a molar, PU 16449, although not necessarily associated, are of the proper size for *Haplolombda*. The canine is more massive than that of CNHM P-15542, but if an allowance for the kind of dimorphism that occurs in *Barylambda* and *Leptolambda* canines is made PU 16449 may not be too large for *Haplolambda quinni*. This specimen is from the Divide Quarry, which is located in the beds of early Silver Coulee age in Big Horn County, Wyoming. From this quarry, an extremely small barylambdid tibia with an associated calcaneum, PU 16481, has also been recovered. It seems probable that this specimen belongs to *Haplolambda*. Except for its smaller size, the tibia resembles that of *Leptolambda schmidti* more than *Barylambda*, and it could possibly be assigned to *Leptolambda*. The calcaneum is definitely barylambdid for it bears the accessory tubercle at the posteromedial base of the sustentacular process which is diagnostic of members of this family.

A fragmentary astragalus, PU 16480, from the same locality as the Princeton University lower dentition of *Haplolambda quinni* No. 16445, completes the list of the specimens which may be referable to this species. It is discussed together with the other pantodont astragali in Section IV. These few bones of the posterior half of the skeleton, which may belong to *Haplolambda quinni*, are mentioned here principally because the post-cranial skeleton associated with the type specimen of *Haplolambda quinni* is of the anterior half of the body only. If the tibia and tarsal elements mentioned above do belong to this species, they show that *Haplolambda* is at least as far along the way toward acquisition of the barylambdid, sub-graviportal stance as are the larger members of its family.

Haplolambda planicanina Flerow, 1952

Archaeolambda planicanina Flerow, 1952, *SSSR Paleontolgicheskii Institut. Trudy* 41: 43–50.

Type: PIN No. 534–68, incomplete lower jaws with left canine through M_3; right P_1–M_3.

Referred specimen: PIN No. 533-133, right maxillary fragment with P^{3-4} and M^1.

Horizon and locality: Early Eocene or latest Paleocene zone, from the vicinity of Ulan and Naran Springs, Nemegetu Basin, Southern Gobi Desert, Mongolia.

Specific diagnosis: Protocone of P^{3-4} having posterior crest broken by a cuspule, analogue of the metaconule ?, size about twenty-five per cent smaller than *Haplolambda quinni*, horizontal rami of mandibles relatively more massive than in the latter species.

Discussion: On the basis of casts and photographs of the type specimens of *"Archaeolambda" planicanina* kindly furnished by Dr. C. C. Flerow, Director of the Paleontological Museum, Academy of Sciences, U.S. S.R., it has been possible to assign this species to the genus *Haplolambda*. Because of this, it is now evident that members of both of the pantodont suborders are distributed in Asia and North America. *Haplolambda planicanina* is the latest surviving member of the Barylambdidae.

LEPTOLAMBDA Patterson and Simons, 1958

Haplolambda sp., Edinger, 1950: 442.
Haplolambda sp., Edinger, 1956: 19, 20.
Leptolambda, Patterson and Simons, 1958: 1–8.

Type species: *Leptolambda schmidti* Patterson and Simons, 1958.

Included species: *Leptolambda schmidti*.

Distribution: Late Paleocene, Tiffanian stage; Silver Coulee beds, Polecat Bench formation, Park and Big Horn counties, Wyoming; Plateau Valley beds, DeBeque formation, Mesa County, Colorado. (The following diagnosis and discussion is abstracted from Patterson and Simons, 1958.)

Diagnosis: I 3/3, C 1/1, P 4/4, M 3/3, no diastema. Incisors above small, with lateral crest on each side; lower incisors larger than upper with lateral crests more prominent, forming a spatulate tooth comparatively longer anteroposteriorly and shorter transversely than in *Barylambda*. Canines smaller than in *Barylambda*, in supposed females incisiform. Premolars of typical barylambdid pattern; P 1/1 one rooted; P^1 usually broader tranversely than in *Barylambda*. Talonid bases of P_{2-3} comparatively larger than in *Barylambda* or *Haplolambda*, with serial homologue of *crista obliqua* directed anteriorly, not running obliquely anterointernally to long axis of tooth as in *Barylambda*. Upper molars differing from those of *Haplolambda* and resembling those of *Barylambda* in M^2 larger than M^1; distinguished from *Barylambda* by presence of heavier cingula on M^{1-2}, absence of backward rotation of protocones of M^{1-3}. Metacone of M^3 less reduced than in *Haplolambda*. Skull with wider, longer frontal area than in *Barylambda* and *Haplolambda*; head as a whole smaller in proportion to body size than in other barylambdids. Cervical centra and neural arches very short, weak; thoracic and lumbar vertebral centra comparatively long. Postcervical vertebrae with lower neural spines than in *Barylambda* and *Haplolambda*. Caudals without (or at most with small, anterior) haemapophyses; caudal centra longer in proportion to width and height than in *Barylambda*. Scapula not as broad anteroposteriorly, particularly in postscapular portion, as in *Barylambda*, broader than in *Haplolambda*; spine proportionately narrower than in *Barylambda*, lacking tuberosity. Clavicle more slender than in *Barylambda*. Bones of fore limb shorter relative to hind than in *Barylambda*; deltopectoral crest of humerus not as flat, less projecting medially; shaft of ulna less broad from front to back, strongly retroflexed, olecranon relatively longer than in *Barylambda* and in *Haplolambda*. Metacarpal I and trapezium not fused as in *Haplolambda*. Metacarpal V much shorter than in *Barylambda*, relatively shorter than in *Haplolambda*, with large lateral expansion in proximal half. Proximal and mesial phalanges differing from *Barylambda* in *much* greater compression (or foreshortening) along axis of digit. Bones of hind limb less robust than in *Barylambda*; femur with more slender, less flattened shaft, third trochanter nearly vestigial; tibia with more slender shaft, no pronounced cnemial crest, distal extremity proportionately less broad; proximal articulation for fibula less projecting laterally; tuber calcis of calcaneum very much shorter.

Leptolambda schmidti Patterson and Simons, 1958

Figures 6, 7, 9, 11, 13

Leptolambda schmidti Patterson and Simons, 1958; *Breviora Mus. Comp. Zool.*, No. 93: 1–8, figs. 1–3.

Type: CNHM No. P-26075, incomplete skull, mandibles, numerous vertebrae and ribs, incomplete scapula and pelvis, various leg and foot bones.

Referred specimens: CNHM Nos. P-26076-7, incomplete skull, various vertebrae and ribs, incomplete scapula, various leg and foot bones; P-15558, P-15571, incomplete dentitions; CM No. 11353, facial region of skull with dentition; PU Nos. 14680, 14879, and 14996, partial skeletons; PU Nos. 14681, 14990, 14992, mandibles; and PU No. 16662, maxilla.

Horizon and locality: DeBeque formation Plateau Valley local fauna, Mesa Co., Colorado; lower levels of the Silver Coulee beds, Polecat Bench formation, Park and Big Horn Counties, Wyoming. CNHM Nos. P-26075-7 from one quarry approximately four miles SSE of DeBeque, Mesa Co., Colorado; CM No. 11353, two to three miles west of DeBeque, Mesa Co., Colorado; PU Nos. 14680 and 14681 from the south side of Polecat Bench, Park Co., Wyoming; PU Nos. 14879, 14990, 14992, and 14996 from separate localities in Big Horn County; PU No. 16662 from Seaboard locality in Sec. 28, T 58 N, R 100 W, Park County, Wyoming.

Diagnosis: As for the genus; size approximately as in *Barylambda faberi* or somewhat smaller, but proportions clearly very different. Specimens from both Colorado and Wyoming fall into two rather sharply defined size groups, one twenty to thirty per cent smaller than the other. The distinction is most apparent in the region of the canine, within the dentition, and in the postcranial skeleton.

Discussion: The ecology and affinities of this animal are treated in some detail by Patterson and Simons, 1958, and will not be taken up again here. Further observations on *Leptolambda* are included in this paper in the comparative sections on pantodont anatomy (IV) and on paleoecology (VI).

A newly discovered left maxilla from the Seaboard Well locality, in Park County, Wyoming, PU 16662, preserves the most complete unworn upper dentition known for this species. On the posterior protocone crest of the P^{2-3} can be seen an accessory cuspule which resembles the cuspule present there in *Haplolambda planicanina*. The dentition and particularly the upper canine alveolus, is small. Hence, PU 16662 is typical of the smaller size division of this species—presumed to be females.

IGNATIOLAMBDA [24] gen. nov.

Type species: *Ignatiolambda barnesi* sp. nov.

Distribution: Tiffanian stage, Tiffany formation, La Plata County, Colorado.

Generic characters: A medium sized pantodont (size of a small ox) with premolar and molar dentition of W-shaped lambdoid type. Canines moderate in size, but relatively large for a barylambdid. Lower canine placed somewhat more labially compared to

[24] *Ignatios*, the Greek form of Ignacio, the town nearest to the locality of this type, plus –*lambda* (λ), in reference to the lambdoid crests on the teeth, and in harmony with the allied genera, *Barylambda*, *Haplolambda*, etc.

other teeth, and angled outward, along with surrounding alveolar border. This angulation causes a distinct bend between the outer face of horizontal ramus and outer face of the symphyseal region below the incisors, instead of a rounded external margin of the ramus as in *Leptolambda*. No pronounced diastema above or below. Lower premolars of barylambdid type; P_1, judging from alveolus, one-rooted; broad talonid on P_{3-4} with *crista obliqua* directed posteriorly, not running from apex to posterolingual angle of tooth as in *Barylambda*. Lower molars with anteroposterior diameter of trigonid slightly greater than talonid, as in all Pantolambdoidea; size increase from M_{1-2} relatively greater than in *Pantolambda cavirictus*. P^{2-4} with typical parametacone V, and fully crescentic protocones having well-developed anterior and posterior cingula, which posteriorly form a distinct shelf. Upper molar paracones and metacones almost equal in size, but paracone slightly larger, basal cingula well developed (particularly the posterior cingulum of the first upper molar), protocone larger and more quadrate than is typical of barylambdids. In this respect most nearly resembling *Haplolambda*. Metacone of M^3 less reduced than in *Haplolambda*. Unguals of manus subequal in size, large, broad, and fissured; not as long relative to width as in *Pantolambda* and *Caenolambda* but much larger; proximal and mesial phalanges about as long as in *Coryphodon*, much longer than in *Haplolambda* and *Leptolambda*.

Discussion: This genus is apparently as distinct from *Barylambda*, *Haplolambda* and *Leptolambda* as they are from one another, but since the one known specimen is poorly preserved, with extensive shattering of the dentition, this form cannot be as well characterized as can the other three. Nevertheless, the diagnostic features given above can be made out with adequate probability, and they show that this specimen has a unique combination of dental structures which make it impossible to refer *Ignatiolambda* to one of the other barylambdid genera without greatly altering present conceptions of these groups.

Ignatiolambda barnesi [25] sp. nov.

Figures 9, 12

Type: AMNH No. 55400, poorly preserved maxilla with C, P^2-M^3 on both sides, horizontal rami of mandible and symphyseal region with C, P_2-M_3 on both sides (several teeth above and below badly shattered). Left manus with four digits preserved and other fragments of anterior part of skeleton.

Horizon and locality: Presumed basal part of Tiffany formation, east of the Florida River and west of Ignacio, north of the Mesa Mountains, SW 1/4, Sec. 29, R 8 W, T 34 N, La Plata County, Colorado.

Specific diagnosis: Not distinguished from generic. Only species of the genus as here defined.

---

[25] Named for Harley Barnes, Jr., who found the specimen.

Discussion: [26] In the illustration of *Ignatiolambda barnesi* (fig. 12) the preserved upper teeth and the left lower dentition are reproduced as they are, together with a restoration of the probable appearance of the left upper and right lower dentitions. Comparison of this restoration and the original specimen with the other barylambdids confirms its affinities with this group, but this specimen shows several interesting distinctions of its own, and a unique combination of other barylambdid characteristics. These differences include the more quadrate form of the upper M^1 and the apparently heavy development of the anteroexternal flange of the jaw ramus, both of which resemble similar features of *Caenolambda*. The significance of this similarity is, however, lessened by the breakage and fracturing of both of these features in the type of *Ignatiolambda* which makes their true conformation uncertain. The lower premolars of *Ignatiolambda barnesi*, on the other hand, differ from these teeth in *Caenolambda* and resemble those of *Leptolambda* in their relatively greater increase in size from front to back. In *Ignatiolambda barnesi* the upper canines are circular in cross section at the base, and show none of the anteroposterior elongation typical of the upper canine of *Caenolambda* and, to a lesser extent, of *Pantolambda cavirictus*.

Like that of *Barylambda*, the I^3 of *Ignatiolambda*, judging by the alveolus, is not separated by a diastema from the canine. The presence of a heavy basal cingulum on the anterior and posterior faces of the protocones of M^{1-3}, combined with a lack of a distinct posterior rotation of the upper molar protocones, as in *Barylambda*, are features approximating the corresponding structures of *Leptolambda*. Altogether *Ignatiolambda* resembles *Barylambda* and *Leptolambda* somewhat more than its does *Haplolambda*, except that the latter genus is more nearly the same size. It is conceivable that if the other barylambdid genera were better understood, or if the type of *Ignatiolambda* were more complete, it might fall within one of the extremes of their morphological range, but in so far as its characteristics can be made out, this new genus of barylambdid appears to be distinct.

The poorly preserved manus of AMNH No. 55400 is informative in one respect in that it indicates that in this species at least four of the unguals were sub-equal in size. There are no other barylambdid feet preserved which so clearly demonstrate this agreement of ungual size in the forefoot. Large and small unguals are associated with several barylambdid skeletons such as CNHM P-14945, P-14902 and PU Nos. 14879, 14996, but it has not been as certain before this find that many of the smaller unguals must belong to the pes, rather than to the lateral phalanges of the manus. Smaller ungual phalanges in the barylambdid pes would

---

[26] I am deeply indebted to Dr. G. G. Simpson of the American Museum of Natural History for making the type of *I. barnesi* available for study and inclusion here. Dr. Simpson also coined the name here given to this pantodont.

agree with the proportionately smaller size of the rest of the metatarsus in these forms. Moreover, the large size of this manus in proportion to the length of the tooth series indicates that *Ignatiolambda*, like the other barylambdids, was comparatively small headed.

The probable geologic horizon of this specimen is discussed in Section V of this paper.

FAMILY **TITANOIDEIDAE**

Type: *Titanoides*, Gidley, 1917.

Included genus: *Titanoides*.

Distribution: Tiffanian stage of the Western United States (Montana, Wyoming, North Dakota, Colorado).

Diagnosis: Middle-sized pantodonts, having full placental dentition. Lower incisors with median cusp, and lateral ridges on each side. Lower canine unique: anteroposteriorly elongated, apex situated anteriorly, triangular at the base of the enamel, and in the different species, showing various degrees of development of a posterior blade (or wing); anterointernal wing also present on canine. P_1 two-rooted; P_{2-4} subquadrate at the base, *crista obliqua* directed posteriorly, talonids increasing in size posteriorly. Molar trigonids more elevated, wider, and longer than talonids; trigonid and talonid V's equal angled. M_{1-3} lacking metastylids, but with large hypoconids, smaller entoconids, both distinct. M_3 with a slight hypoconulid crest. Upper incisors smaller than lower incisors, increasing slightly in size from I^1 to I^3. Upper canine very large, flattened transversely, elongated and sabre-like; short diastemata anterior and posterior to upper canine. Primitive pantolambdid crescents retained in premolar and molar series. P^1 two-rooted; protocones of P^{2-4} placed somewhat anterior to transverse midline of tooth. M^{1-2} basically triangular in outline, relatively broader anteroposteriorly and narrower transversely than in other pantodonts; V's of paracone and metacone subequal, parastyle and metastyle equally extended labially. M^3 parastyle prominent, metacone and metastyle reduced.

No posterior extension of angle of jaw, short coronoid process; no fusion of symphysis. Skull with prominent sagittal crest—lacking the flattening of skull roof characteristic of *Coryphodon*. Premaxillaries relatively small, with slender ascending rami; zygomatic arches slender, and bowed outward from skull. Head large in proportion to body size. Cervical vertebrae comparatively large, with strong neural arches; centra of thoracic and lumbar vertebrae relatively long. Sacrum with low spines, transverse processes fused distally and articulating with pelvis; anterior caudals small and flattened, tail apparently short.

Clavicle wide and flattened, not elongated; coracoid process of scapula well developed. Humerus with extensive articular surface on head and great trochanter, deltoid crest prominent, entepicondylar foramen present. Manus unique: with fusion of scaphoid and centrale, and with high, rounded lunar showing little lateral

expansion. Phalanges moderately long with extensive proximal and distal articular surfaces, having rugosities for muscle attachment, and at least on the first three digits bearing large, laterally compressed claws.

Pelvis moderately expanded; femur with third trochanter small and placed high on the shaft; tibia with prominent cnemial crest, intermediate in development between that of the Pantolambdidae and the Coryphodontidae.

Tarsus not particularly distinctive (phalanges unknown); astragalar foramen present: tuber calcis relatively slender, lacking an accessory internal tubercle.

Discussion: The history of taxonomic assignments of this particular group of pantodonts has, since the description of the type species in 1917, followed a rather complex course. Gidley (1917: 435) considered this form to be related to the titanotheres (hence the name *Titanoides*), but noted as well its amblypod (pantodont) affinities. He did not attempt to assign *Titanoides* to a separate family on the basis of the limited material available to him, but suggested that future discoveries might make this necessary. Jepsen (1930: 508) reports that "Mr. Gidley has long since regarded *Titanoides* as an amblypod."

After Gidley's publication, no further material referable to *Titanoides* appeared until Jepsen (1930: 506) described a partial lower dentition of a new species, *Titanoides gidleyi*, from the Silver Coulee beds of the Polecat Bench formation, Bighorn Basin, Wyoming. Jepsen's suggestion that *Titanoides* might represent a direct ancestor of *Coryphodon* was a most reasonable conclusion considering the incomplete dentitions of *Titanoides* then known. Probably the most pronounced differences between these two genera, in the parts of jaw and dentition available for comparison in 1930, were the greater development of the premolar and molar paraconids, and the higher elevation of the *crista obliqua* in the molar series of *Titanoides*. Because of the many important similarities which he observed between the lower dentitions of *Coryphodon* and *Titanoides*, Jepsen placed *Titanoides* in the family Coryphodontidae.

In 1929 a new type of pantodont was discovered in beds of Paleocene age in Mesa County, Colorado, and further remains of this form were collected in 1933 by an expedition of the Chicago Natural History Museum under the direction of Bryan Patterson. Patterson (1933: 417) assigned the newly discovered form to the genus *Titanoides*, but made it a separate species, *Titanoides faberi*, and in the next year he proposed a new subfamily Titanoidinae (1934: 72) to include the genus *Titanoides*. Patterson later concluded (1937: 229) that this initial assignment of the species *faberi* to the genus *Titanoides* was not justified, and consequently proposed a new genus for the Colorado species, calling it *Barylambda faberi*. At the same time he created the subfamily Barylambdinae to include this species, and to this subfamily he provisionally referred *Titanoides.*

Later in the same year Scott, recognizing the distinctive features of *Barylambda faberi,* placed both *Barylambda* and *Titanoides* in a new family, Titanoideidae (1937b: 479). Since, however, it had not been realized what distinctively different types of pantodonts *Barylambda faberi* and *Titanoides primaevus* were, and no upper dentition or post-cranial skeleton was known for *Titanoides,* Scott's discussion of the Titanoideidae was based almost entirely on *Barylambda faberi.* The latter form, as will be pointed out below, was subsequently removed by Patterson to its own family, so that what Scott actually recognized in 1937 was that *Barylambda* represented a distinct family. Removal of this genus from the Titanoideidae left this family almost entirely without definition. The few distinct characters of the lower dentition of *Titanoides primaevus* then known did not indicate a need to maintain a separate family for this form.

Simpson (1937b: 14) made the following observation about the taxonomic position of *Titanoides* at that time.

In removing *T. faberi* from *Titanoides* to a new genus, *Barylambda,* Patterson (1937) has necessarily deleted the subfamily Titanoidinae, since the characters assigned to that group are not in fact known in *Titanoides,* and substituted Barylambdinae. He now refers *Titanoides* (*sensu stricto*) to that subfamily, but strongly emphasizes the tentative nature of the reference. It cannot be determined until skeletal material of *Titanoides* is found.

Also in 1937, in a revision of the Paleocene faunas of the Crazy Mountain Field, Montana, Simpson (1937a: 265) presented a detailed analysis of the fallacies involved in associating the groups now designated as the Dinocerata and Pantodonta in a single order, the Amblypoda, and raised the two suborders to ordinal rank. He recognized two families of pantodonts, the Pantolambdidae and Coryphodontidae, but did not specify to which of these families the genus *Titanoides* might be referred. In the same year Simpson (1937b: 11) described a new pantodont species, *Titanoides zeuxis,* based on a fragmentary lower dentition from the Crazy Mountain Field, Montana. In discussing this form he stressed that only a very tentative assignment to family could be made for *Titanoides.* With reservations he placed this genus in the subfamily Pantolambdinae.

In the course of later expeditions by the Chicago Museum of Natural History to the Plateau Valley area of Mesa County, Colorado, in the late 1930's many more specimens of *Barylambda faberi* were collected, so that at present it is one of the best known Paleocene mammals. Two new species, representing distinctly different types of pantodonts, were discovered as well. These two new forms, *Sparactolambda looki* and *Haplolambda quinni,* were described by Patterson in 1939, along with further observations on *Barylambda,* and at that time he raised the Barylambdinae to family

status, including in this new family the genera *Barylambda* and *Haplolambda.* All other pantodont species then known, with the exception of those belonging to the dubiously placed Mongolian Pantolambdodontidae, which he did not attempt to assign taxonomically, were referred by him to the Coryphodontidae. Neither did he try, at that time, to divide the Coryphodontidae into subfamilies but suggested that future discoveries might clarify the taxonomic relationships of this heterogeneous group of pantodonts.

In summary, it should be pointed out that by 1939 the genus *Titanoides* had been successively classified as follows: first with the titanotheres by Gidley (1917); then placed in the Coryphodontidae by Jepsen (1930); later in a distinct subfamily of the coryphodontids, Titanoidinae (Patterson, 1934); subsequently in a new family Titanoididae (Scott, 1937); in the same year in the Pantolambdinae (Simpson, 1937b); and finally reassigned to the Coryphodontidae (Patterson, 1939). The prolonged uncertainty as to familial assignment for *Titanoides* was largely a result of the problems encountered in trying to classify such a fragmentary type. The two forms described subsequently to the type species, *Titanoides gidleyi* and *Titanoides zeuxis,* were also based on specimens that were difficult to classify, consisting only of fragmentary lower dentitions.

After 1939, pending discovery of new specimens, no reassignment of *Titanoides* was attempted. In 1951, however, a rather surprising find by a Princeton University field party made possible a much better understanding of this entire taxonomic problem. Since the initial publication on *Titanoides primaevus* by Gidley, many persons have visited the locality, erroneously indicated by a photograph in Gidley's paper as being that of the type specimen, without finding any further parts. Dr. Vernon Bailey, who collected the type of *Titanoides primaevus* in July, 1913, about three miles northeast of old Fort Union (now Buford), North Dakota, in the type section of the Paleocene Fort Union formation, believed that more of the specimen might be found at the spot of the original discovery, and with this in mind he gave two photographs of the correct locality to Dr. Glenn L. Jepsen of Princeton University. In 1951, with the help of these photographs, Mr. Robert Witter, then Preparator in Vertebrate Paleontology at Princeton, was able to relocate the spot where the type lower dentition had been collected and there, with Mr. Frank N. Goto, recovered a complete upper right dentition of *Titanoides,* as well as the left and right petrosals, occipital condyles, and other tooth and bone fragments. The photographic identification of the spot, as well as similarity of matrix and type of weathering of the teeth and bone in each of these finds at the locality from which the type of *Titanoides primaevus* was discovered, makes it virtually certain that they belong to the same individual, although the two discoveries were made thirty-eight years apart.

As a result of comparisons of an upper right molar

series of a pantodont from the Bison Basin, Fremont County, Wyoming (U. of Wyoming No. 1039) with the newly discovered upper dentition of the type *Titanoides primaevus*, and with other pantodont species, Gazin (1956: 47) concluded that the genus *Sparactolambda*, Patterson (1939a) is synonymous with *Titanoides*. It now seems reasonably certain that the species *Sparactolambda looki* is also a synonym of *Titanoides primaevus*. The differences between the upper dentition of the type of *Titanoides primaevus* and that of *Sparactolambda looki* are in fact remarkably slight. The resemblance is still more striking when one considers that it is highly improbable these specimens from Wyoming and Colorado could have been exactly contemporaneous. Most populations of single species of large living mammals as separate geographically as these areas of North Dakota and Colorado could probably be shown to exhibit greater variation in dental characters in spite of the fact that they are coeval. Merriams' classic fractionation of the American populations of *Ursus* into more than eighty species and subspecies is an example of the errors which a too zealous search for minor differences can produce.

The identity of *Titanoides primaevus* and *Sparactolambda looki* has now made it possible to detail the skeletal characters which distinguish this quite unique group, the titanoideids, as a separate family; for the skeletal material from the Plateau Valley fauna, formerly referred to *Sparactolambda looki*, includes the skull, mandible, much of the spinal column, all of the limb bones, pectoral and pelvic elements, and most of the elements of the pes. The conclusion that *Titanoides* represents a distinct family is discussed in more detail in Section IV on the skeletal anatomy of the Pantodonta, and in the discussions of the major subdivisions of the order at the beginning of the present section.

It should be pointed out here, however, that the many similarities between *Titanoides* and *Coryphodon* which Patterson (1939a) observed, suggest that the two forms might have shared a common ancestor subsequent to the splitting apart of the other pantodont stocks. These points of resemblance, nevertheless, are largely confined to the spinal column, pelvis, and some of the limb bones. It would be difficult to show that these agreements in skeletal conformation are not due to the common pantodont heritage of the two forms, as well as to parallelism resulting from increase in size.

Skeletal similaries in many of the archaic mammalian orders, for instance those between *Ectoconus* and *Pantolambda bathmodon*, which have been effectively detailed by Matthew (1937: 164), are pronounced. One should probably not rely too heavily on such similarities when considering affinities between forms as different as *Titanoides* and *Coryphodon* as long as the conventional procedure is to consider the Pantodonta, Condylarthra, Dinocerata, etc., separate orders in spite of the many broad resemblances in the skeletons of these archaic mammals.

In the dentition, basicranium, and structure of the pes, *Coryphodon* and *Titanoides* are as divergent as are any of the Pantodonta.

TITANOIDES Gidley, 1917

Titanoides Gidley, 1917: 431.
Sparactolambda Patterson, 1939: 352.

Type: *Titanoides primaevus* Gidley, 1917.

Included species: *Titanoides primaevus, T. gidleyi, T. zeuxis, T. majus, T. simpsoni.*

Distribution: Torrejonian stage: Lebo No. 2, Crazy Mountain Field, Sweetgrass County, Montana (Gidley Quarry). Tiffanian stage: Silver Coulee beds, Polecat Bench formation, Park and Big Horn Counties, Wyoming (Crocodile Tooth Quarry, Cedar Point Quarry, and other localities); Melville formation, Crazy Mountain Field, Sweetgrass County, Montana (Douglass Quarry, and other localities); "Titanoides," Ledge, and West-end localities, Bison Basin, Fremont County, Wyoming; Fort Union formation (type section), Sentinel Butte shale member, Williams and McKenzie Counties, North Dakota; Plateau Valley local fauna, DeBeque formation, Mesa County, Colorado.

Diagnosis: As for the family.

Discussion: The five species listed above apparently represent distinct forms, but the known differences between them, with the exception of *Titanoides simpsoni*, although sufficient to suggest that each species is a distinctly adapted population, are not great. It has been a frequent occurrence in this order, however, that slight differences in the dentitions are associated with more radical differences in the rest of the skeleton. In other words, the dentition of most pantodonts remains generalized, or unspecialized while the remainder of the skeleton, particularly its extremities, exhibits radically different adaptations. If this relationship of the pantodonts in general, between little-specialized dentitions and more specialized post-cranial anatomy, is also true of species included in the genus *Titanoides*, then it might be expected that, when more of the post-cranial skeleton of these species is known, this material will emphasize the less pronounced but significant differences of the dentitions of these five species.

Titanoides primaevus Gidley, 1917

Figures 1, 3, 4, 5, 6, 7, 10, 14, 15

Titanoides primaevus Gidley, 1917, *Proc. U. S. Nat. Mus.*, 431–435, 1 fig., 1 pl.
Sparactolambda looki Patterson, 1939, *Geol. Ser. Field Mus. Nat. Hist.* 6: 354–361, figs. 100–103.

Type: USNM No. 7934, lower right P$_3$, M$_1$ (trigonid), M$_2$ and M$_3$; lower left M$_2$ (posterior part of tooth), left and right parts of ventral region of the symphysis. Presumed remainder of type, PU No. 16490, upper right premaxilla with I^1 to I^3, upper right maxilla with C to M^3, fragmentary upper left teeth from P^2 to M^3; both petrosals and occipital condyles, other tooth and bone fragments.

Referred specimens: USNM No. 20029; UW No. 1093; CNHM Nos. P-15520, P-15523, P-14984, P-15547, P-M 238, P-M 240, P-M 246; PU Nos. 16490 and 16448.

Horizon and locality: Fort Union formation, three miles NE of Buford, Williams County, North Dakota, NW 1/4 Sec. 29, T 153 N, R 102 W, McKenzie County, North Dakota; DeBeque formation, Plateau Valley fauna (several localities), Mesa County, Colorado; "*Titanoides*" locality, Bison Basin, Fremont County, Wyoming; Silver Coulee beds, Polecat Bench formation, Park County, Wyoming, Center of Sec. 2, T 56 N, R 99 W.

Diagnosis: Patterson (1939a: 352) has already published a description of this species under the name of "*Sparactolambda looki*." The present diagnosis is intended to detail those features of the dentition which have been found useful in comparing *Titanoides primaevus* with the other species of *Titanoides*.

I 1/1 smallest, I 2/2, I 3/3 subequal in size; upper incisors consisting of transversely flattened cones, lower incisors having central cones with lateral cusps (or wings) lengthening the anteroposterior diameter of the teeth; lower incisors much larger than upper. Lower canine greatly elongated by the development of a posterior shearing shelf (or blade), apex recurved posteriorly and hooklike, situated at anteroexternal angle of tooth; anterointernal wing well developed. P_1 two-rooted, paraconid of P_{2-4} much smaller than metaconid, particularly on P_2; talonids of P_{2-4} somewhat lower than molar talonids, slight development of the internal basal cingula. M_{1-3} with trigonids longer and wider than talonids; entoconids consisting of distinct, small cuspules. Upper canine elongated with longitudinal ridges in the enamel converging at the apex, bean-shaped in cross section at the base; having three faces, the internal wearing against the lower canine, the external side of the tooth divided into an anterior and posterior face. P^{1-4} successively increasing in size from front to back, P^1 with an internal cingulum in the position of protocone, external wings not extended transversely from the base of the parametacone as in the succeeding teeth; P^{2-4} protocones shifted anteriorly, posterior cingulum (position of hypocone) forming a heavier shelf than anterior cingulum. M^{1-2} roughly triangular, parastyle and metastyle extended labially and also producing a relatively greater anteroposterior diameter, paracone and metacone V's subequal but paracone slightly more internal; protocone broadened anteroposteriorly but relatively little expanded lingually; slight development of protoconule and metaconule. M^3 parastyle usually projecting labially to a greater degree than in other pantodonts; metacone reduced, metastyle absent. M^{1-3}, anterior and posterior cingula of protocone occasionally

united across lingual base of protocone, but never forming a cingular shelf.

The cranial and post-cranial anatomy of this species is that on which the family Titanoideidae is based and it has already been outlined in the discussion of this family. More detailed consideration of the post-cranial distinctions of *Titanoides primaevus* is included in Section IV.

Discussion: Although there is little doubt that the upper dentition of *Titanoides primaevus* from the site of the type specimen belongs to the same individual as the type, USNM 7934, there are no actual contacts between the parts preserved in the two finds. The Princeton specimen has consequently been given a separate number, PU 16490. Nevertheless, PU 16490 is referred to here as the upper dentition of the type of the species. One of the principal reasons why the specific identity of the type lower dentition of *Titanoides primaevus* with *Sparactolambda looki* was not recognized before the upper dentition of the type of *Titanoides primaevus* was known is that the right and left symphyseal fragments associated with the type lower teeth appeared to have alveoli of premolar roots penetrating to within about one-half centimeter of the ventral surface of the mandible. *Titanoides primaevus* was consequently considered to have had a shallow horizontal ramus, entirely distinct from that of *Sparactolambda looki*, which has a deep ramus below the premolar series with the roots ending about three centimeters above the lower edge of the jaw. These supposed alveoli in the type specimen, however, are not arranged with their greatest length transverse to the ramus, as would be expected, but are asymmetrically placed, with the greatest lengths directed at oblique angles to the main axis of the ramus. Further, these cavities resemble alveoli only in the right symphyseal fragment and are much less distinct on the left side. Comparison of this region with other broken mandibles of *Titanoides*, particularly with the type of *Titanoides majus*, demonstrates that cavities in the bone often occur in this genus at a distance considerably below the actual alveoli of the premolar roots. These cavities certainly are not alveoli of roots, and they consequently do not indicate that the type of *Titanoides primaevus* had a shallow jaw. Instead, the openings probably were connected with the termination of the nutrient canal of each horizontal ramus, which does not extend beyond the huge root of the lower canine. X-rays of the mandible of the type of *Sparactolambda looki* show that cavities of this sort are also present in this specimen. If one accepts such an interpretation of the cavities in the mandibular fragments of *Titanoides primaevus*, there is little else to distinguish the lower dentition of *Sparactolambda looki* from the comparable teeth preserved in the type of *Titanoides primaevus* except that the type is somewhat larger in most dimensions and possesses a slightly higher *crista obliqua* and entoconid than the DeBeque specimens. The latter distinctions do not appear to justify separating the two forms specifically, nor are there any distinctions between the upper dentition of the type of *Titanoides primaevus*, PU 16490, and that of *Sparactolambda looki*, which would indicate that they are different species.

The most striking features of the dentition that have been useful in distinguishing *Titanoides primaevus* from the other species of *Titanoides* are the anterior shifting of the protocones of the upper premolar series, P^{2-4}, and the relatively smaller paraconids of the lower premolars and molars. It seems possible that these two structures interact functionally and consequently are mutually influencing. Another distinctive feature of this species is that the transverse diameter of the third molar above usually approximates that of the M^2, rather than being considerably greater. The specimen from the Bison Basin, UW 1093, considered by Gazin (1956: 47) to belong to this species, does appear to be *Titanoides primaevus*, but this assignment is based mainly on size. In the absence of the anterior dentition above and below, it is difficult to assign specimens to a particular one of the four species of *Titanoides* except by size, a procedure which is usually not satisfactory. The most diagnostic dental feature of *Titanoides primaevus* is the elongate lower canine; but, as this is preserved in only one of the known dentitions, it has not been helpful in distinguishing members of this species. Because the distinctive features are so rarely preserved, it is not entirely certain whether *Titanoides primaevus* occurs in the collections of pantodonts from the Bighorn Basin of Wyoming.

One specimen, however, from the lower Silver Coulee beds of the Polecat Bench formation, PU 16446, a partial upper right dentition, can be provisionally referred to *Titanoides primaevus*. The teeth included in this specimen are both P^3's, a lingual fragment of the right P^4, and the right M^{1-3}. Although these teeth average more than ten per cent smaller than those of the type in corresponding measurements, they agree with *Titanoides primaevus* in having the protocones of the P^{3-4} shifted anteriorly. The M^3 also agrees with the type of *Titanoides primaevus* in being relatively smaller in proportion to the M^2 than in *Titanoides gidleyi*.

Titanoides zeuxis Simpson, 1937

Titanoides zeuxis Simpson, 1937; *Amer. Mus. Nov.*, No. 940: 11–15, fig. 4.

Type: AMNH No. 35201, left lower jaw with P_3, trigonids of P_4, M_2, M_3 and other fragments.

Referred specimens: CNHM No. P-15551, and PU No. 14617.

Horizon and locality: Melville formation, Crazy Mountain Field, Montana, about 1,000 feet above Scarritt Quarry level, Sec. 26, R 14 E, T 5 N, Sweetgrass County, Montana, and from the Douglass Quarry, Sec. 18, T 5 N, R 15 E, Sweetgrass County, Montana. Plateau Valley local fauna, DeBeque formation, Mesa County, Colorado, one mile north of D. Harris ranch.

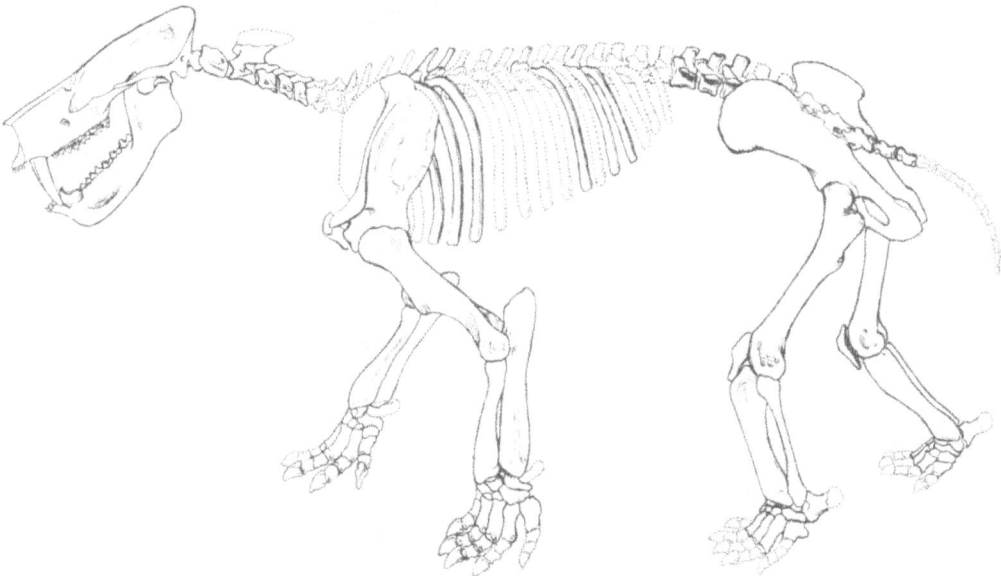

FIG. 3.   × 1/10. *Titanoides primaevus* Gidley. Based principally on CNHM P-15520, with details of cranium added from PU 16490 and USNM 20029.

Specific characters: Characters are given in Simpson (1937b: 11) but the new specimens from Colorado make it advisable to redefine the species as follows:

Smallest of the Tiffanian species of *Titanoides*, jaw equivalent in size to that of *Haplolambda quinni*, but in other respects conforming in proportion to *Titanoides primaevus*; ranging from twenty to fifty per cent smaller than the latter species in corresponding measurements of the lower dentition. Resembles *Titanoides primaevus*, CNHM P-15520, and differs from *Titanoides gidleyi* in the greater reduction of the paraconids relative to the metaconids of the lower premolar and molar series. Heel of M_3 large, with relatively distinct entoconid. Hypoconulid crest present on talonid of M_2. Lower canine with an anterior cingulum, and much smaller relative to tooth series than in *Titanoides primaevus* but with pronounced anterointernal wing and posterior blade or wing, as in that species.

Discussion: Simpson suggested (1937b: 13) that the lower cheek teeth of this form are in several respects structurally intermediate between those of *Titanoides primaevus* and *Pantolambda*. One of the characters which he thus considered intermediate was the possible presence of a hypoconulid on the M_3 of the type specimen, AMNH 35201, but erosion and breakage of the talonid makes the presence of this feature uncertain. No hypoconulid is present on the M_3 of the Plateau Valley specimen, but a crest in this region is typical of *Titanoides*. A further character of this form which Simpson mentions as intermediate between the corresponding structures in *Titanoides primaevus* and *Pantolambda*, namely, "relatively low molar talonids," is a conclusion which seems to be based on the misidentification of a trigonid, considered to be that of the M_1. Comparison of this trigonid with those of other specimens of *Titanoides* not known at the time of description of this species shows that the trigonid almost certainly belongs to the P_4. Interpreted as part of a P_4, this fragment is not unusual, for the low premolar talonid present here is typical of *Titanoides*, in which the premolar talonids are always lower than those of the molars. The molar series of this species consequently shows no real approach to those of *Pantolambda* in this respect. As far as determinable, *Titanoides zeuxis* possesses all the characters of the dentition that have been found useful for distinguishing the titanoideids from other pantodonts. This species does not exhibit an intermediacy which would necessitate an arbitrary separation of the titanoideids and pantolambdids. The lower canine of the Colorado specimen of *Titanoides zeuxis* is as distinct from the lower canine of *Pantolambda* as is that of any pantodont. On the other hand, in many of its features it is like the lower canine of *Titanoides primaevus*. In spite of the very fragmentary specimens available in 1937, Simpson's final conclusion about *Titanoides zeuxis* is essentially correct: "in parts actually known now, *Titanoides* seems to me

somewhat closer to *Pantolambda* than *Barylambda*" (1937b: 14).

The referred specimens of *Titanoides zeuxis* from the DeBeque formation, CNHM 15551, are useful in clarifying the affinities of this species. These partial lower dentitions of two adult individuals from the same locality include from one side or the other the C_1, P_2, P_4, M_{1-3}, most of the horizontal ramus of the jaw of one individual, and the symphyseal regions of both, one upper right premolar and several phalanges including one well preserved ungual phalanx.

The DeBeque mandibular fragments agree with those of other species of *Titanoides* in lacking fusion of the symphysis; in the position of the mental foramen beneath the canine, in having the greatest depth of the horizontal ramus beneath the canine, and the least at the posterior margin of the M_3; and in possessing moderately developed flanges along the anteroexternal margins of the horizontal rami. The left lower canine bears an anterior cingulum not present in *Titanoides gidleyi*, but the presence or absence of this feature cannot be determined in the DeBeque dentition of *Titanoides primaevus*, CNHM P-15520, since the lower canine of that specimen is not fully erupted. In *Titanoides zeuxis* the lower canine also possesses an anterointernal wing from which a prominent ridge descends along the internal side of the tooth root. The presence of this wing and ridge, coupled with the anteroposterior elongation of the tooth, gives it a triangular cross section at the base of the enamel and throughout the length of the root. Such a conformation of the canine is diagnostic of the titanoideids. It is not known to occur in the pantolambdids, barylambdids, or coryphodontids (*Coryphodon*).

These characters, together with the strongly developed posterior blade of the lower canine, suggest that *Titanoides zeuxis*, like the other titanoideids, possessed greatly elongated upper canines and the general digging and rooting adaptation which has been suggested for the genus (see Patterson, 1939a: 354). Corroboration of this possibility is afforded by the elements of the pes found in association with the two DeBeque dentitions of *Titanoides zeuxis*. These consist of five mesial or proximal phalanges and one claw, which are all approximately twenty-five per cent smaller than comparable phalanges of *Titanoides primaevus*. The presence of claws in this species, as in *Titanoides primaevus*, rather than the more typical fissured unguals of the pantolambdids and barylambdids, tends to confirm the view that a clawed manus is typical of the titanoideids, and in conjunction with other skeletal modifications, to suggest a chalicothere-like adaptation.

A specimen in the Princeton collection, No. 14617, from the Melville formation of the Crazy Mountain Field, most probably belongs to *Titanoides zeuxis*. Since, however, it is from the Douglass Quarry, which is about 1,500 feet lower in the Melville than the locality of the type of *Titanoides zeuxis*, some question as to specific identity of the two finds might be maintained. PU 14617 consists of a fragment of a mandibular ramus including the P_{1-2}. These teeth agree with *Titanoides zeuxis* in general size, but since the P_1 is not known in other specimens of *Titanoides zeuxis*, comparisons are possible only between the lower P_2's. The P_2 of the Douglass Quarry specimen differs from the type of *Titanoides zeuxis* in having a greater anteroposterior diameter, produced by a more marked anterior extension of the cingulum at the base of the paraconid; but in size and most other respects the two teeth are quite similar. They agree, for instance, in having subequal development of the paraconid and metaconid. One difference is that the Princeton specimen is widest on a transverse line through the metaconid, while the P_2 of CNHM 15551 is widest on a transverse line through the paraconid. PU 14617 agrees with *Titanoides primaevus* in lacking a ridge on the anterolingual slope of the paraconid, but differs in possessing a relatively less developed metaconid, which, in the latter species, is very large and is shifted forward to a central position on the lingual margin of the tooth. In both *Titanoides primaevus* and the Douglass Quarry specimen the P_1 is much narrower transversely than the P_2; in other respects the P_1 in both forms is quite similar although the above mentioned differences in the P_2 can be used to distinguish *Titanoides zeuxis* from *Titanoides primaevus*.

Two isolated lower right canines from Crocodile Tooth Quarry in Big Horn County, Wyoming, PU Nos. 16462 and 16471, are close in size and proportion to the canine of the DeBeque specimen of *Titanoides zeuxis*, CNHM P-15551, but they have a somewhat shorter posterior blade. It is not possible to demonstrate conclusively that these are not milk canines of the larger *Titanoides gidleyi*, which occurs in the same quarry. If, however, these two canine teeth do pertain to *Titanoides zeuxis*, the occurrence of the two species in this quarry confirms their contemporaneity.

Titanoides gidleyi Jepsen, 1930: 506

Figure 4

Type: PU No. 13235, fragmentary lower jaws of one individual, with all teeth represented on one side or the other, except I_1, I_2, P_1 and talonid of M_3.

Distribution: Silver Coulee beds, Polecat Bench formation, Park and Big Horn Counties, Wyoming.

Referred specimens: PU No. 14974, part of a left maxilla with P^{2-4}, M^{1-3} and PU Nos. 14976, 16451, 16453, 16459, and 16448.

Diagnosis: A more complete diagnosis of the type lower dentition is included in Jepsen's original description of the species. The features summarized here are those which appear to have the greatest taxonomic value in comparing this form with the other species of *Titanoides*.

Dental formula, I ?3/?3, C 1/1, P 4/4, M 3/3. Lower incisors flattened transversely with small accessory cusps on the sides of a larger median cusp; lower canine having an anterointernal wing, supported by a distinct vertical ridge on the internal side of the root; posterior blade of the canine little developed. Metaconids of lower premolars relatively smaller than in *Titanoides primaevus*. Third molar, above and below, much larger, relative to the rest of the molar series, than in *Titanoides primaevus*. Corresponding measurements on the teeth averaging about ten per cent smaller than in the type species.

Discussion: The only associated pantodont dental series from Crocodile Tooth Quarry [27] is preserved in a left maxilla, PU 14974, which almost certainly belongs to *Titanoides gidleyi*. Several isolated lower teeth from the same quarry, including a left lower canine, PU 16453, can also be referred with reasonable certainty to *Titanoides gidleyi*. The absence of any teeth definitely assignable to *Titanoides primaevus* further suggests that other isolated titanoideid teeth from this locality, most of which cannot be distinguished as to species, also belong to *Titanoides gidleyi*. The upper dentition of *Titanoides gidleyi* differs from that of the type species in having a more centrally placed protocone on the P^3, a much larger M^3 relative to the M^2, and in the posterior cingulum of M^3 forming a larger shelf than it does in *Titanoides primaevus*. In *Titanoides gidleyi* the transverse diameter of the M^3 is one-third larger than that of the M^1, while in *Titanoides primaevus* the corresponding measurement is only one-fifth larger.

It is therefore evident that the species *Titanoides primaevus* and *Titanoides gidleyi* do not differ markedly except in absolute size and relative proportions of some canine, premolar, and molar structures; these one might attribute to individual variation. Closer examination, however, of the probable adaptive functions of the dentitions of these two species does not readily support this view. The most significant difference in the two forms is in the shape of the posterior lobe or blade of the lower canine, which in *Titanoides primaevus*, CNHM P-15520, is greatly enlarged, so as to form a shearing shelf for the posterointernal face of the upper canine. The lower canine of *Titanoides gidleyi*, while possessing a reasonably

[27] The type specimen of *T. gidleyi* was found in Silver Coulee beds of the Polecat Bench formation at a somewhat lower level than the Princeton Quarry. The new material here referred to *T. gidleyi* is also of Silver Coulee age, judged on a basis of the faunal correlations of mammalian species from several quarries discovered by Princeton University field parties at approximately the same stratigraphic levels. These quarries are located to the southeast of Polecat Bench proper, in Big Horn County, Wyoming, on a NE–SW line between Lovell and Greybull, Wyoming, in T 54 and 55 N, R 95 and 96 W. Two of these quarries, Crocodile Tooth Quarry and Cedar Point Quarry, have yielded teeth and jaw fragments of *T. gidleyi*.

similar triangular shape in cross section at the base, does not have the posterior lobe enlarged. In fact, the lower canines of *Titanoides gidleyi* are somewhat more similar in appearance to those of *Coryphodon* than they are to those of *Titanoides primaevus*. It does not seem likely that what appears to be such an efficient dental mechanism as the lower canine of *Titanoides primaevus* could be a dimorphic or variable character. The absence of a pronounced posterior canine blade in *Titanoides gidleyi* can consequently be taken to indicate that feeding habits and ecology were different in the two forms and that they most probably do represent distinct species. The parallelism exhibited between this shearing canine blade and the greatly elongated P_1 in *Caenolambda jepseni*, which performs a similar shearing action against the posterointernal face of the upper canine, suggests that this type of slicing mechanism has some particularly useful adaptive function in the pantodont phylum. That such a sectorial region should have evolved twice, but incorporating different teeth to accomplish a similar mechanical function in the same region of the jaw, further decreases the likelihood that either of these two arrangements are only individual variations.

Titanoides majus sp. nov.[28]

Figure 18

Type: PU No. 16447, parts of right jaw ramus with symphyseal region, P_2, P_3, part of P_4, and roots, or alveoli, of canine, P_1, and M_1.

Distribution: Lowermost Silver Coulee beds near Sage Point, Polecat Bench formation, Park County, Wyoming.

Diagnosis: Dental formula below, ?3-1-4-3 diastema between canine and P_1; P_1 two-rooted, slight diastema behind P_1, no apparent anteroposterior elongation of canine root; heavy symphysis and flare at anteroexternal margin of horizontal ramus; metaconids of lower P_1 and P_2 relatively smaller than in *Titanoides primaevus*; paraconid and metaconid of P_2 farther apart, and P_3 more quadrate at the base than in either *Titanoides gidleyi* or *Titanoides primaevus*; size distinctly larger in known parts than both *Titanoides primaevus* and *Titanoides gidleyi*.

Discussion: This very fragmentary specimen from one of the lowest strata of the Silver Coulee beds was recovered at Sage Point on the northeast side of the Polecat Bench proper, not far above the Rock Bench (Torrejonian) level. It is consequently of great interest, since PU 16447 is probably the oldest known specimen of *Titanoides* of Tiffanian age.

If it can be safely assumed that the pantolambdids, at least in some of their more generalized features,

most nearly approximate the ancestral pantodont stock, then it might be expected that *Titanoides majus*, being the earliest known species of this genus in which a part of the lower dentition is preserved, would show a few more primitive features than other titanoideid species, and that these features would resemble pantolambdid structures. To a limited extent this appears to be true of the lower canine of *Titanoides majus*. The crown of the canine is missing, but the tip of the root and part of the walls of the canine alveolus remain in the jaw fragments of this specimen. These fragments show that the canine could not have been much elongated anteroposteriorly, and that in the preserved root tip the tooth is sub-circular in cross section. The premolars, however, are like those of the other species of *Titanoides* in general conformation, so that there is little likelihood that this form could be referred to the Pantolambdidae. Except for absolute size and some details of proportion, the P_{1-2} of *Titanoides majus* are much like those of *Titanoides gidleyi*. Possibly the rounded lower canine and diastemata before and after the lower P_1 are indicative of the first stage in an ascending structural series leading from *Titanoides majus*, through *Titanoides gidleyi* to *Titanoides primaevus* and detailing the evolutionary development of the pronounced posterior blade of the lower canine of *Titanoides primaevus*. Apparently there is no posterior canine blade in *Titanoides majus*, judging from the rounded curvature of the alveolar wall in this region, and there are diastemata following the canine and P_1. In *Titanoides primaevus*, on the other hand, the posterior extension of the canine is so great that the premolars are crowded against one another, and the anterior root of the P_1 actually overlaps the root of the canine on its posteroexternal margin. The canine of *Titanoides gidleyi* is intermediate between the other two species with regard to development of the posterior blade and crowding of the premolars.

Titanoides simpsoni[29] sp. nov.

Figure 10

Type: AMNH No. 35720. Right maxilla with P^4, M^{1-3}.

Horizon and locality: Torrejonian stage. Gidley Quarry NW 1/4 of NE 1/4, Sec. 23, T 5 N, R 15 E. About 1,200 feet above the base of Lebo formation no. 1 Crazy Mountain Field, Sweetgrass County, Montana.

Specific diagnosis: Smallest of the species of *Titanoides*, intermediate in size between *Pantolambda bathmodon* and *Pantolambda cavirictus*. Resembling other species of *Titanoides* and differing from pantolambdids in more triangular form of M^{1-2}, in possession of a lingual cingulum on upper molar protocones of *Titan-*

---

[28] From the Latin *majus*, larger, in reference to the size of this species compared with the other known members of the genus.

[29] Named in honor of Dr. George G. Simpson, under whose direction this species was collected, and in appreciation for his major contributions to the understanding of the Paleocene mammalian faunas of North America.

oides type, not a cingular shelf; in greater labial extension of metastyle of M^{1-2}; in more distinct crenulation of surfaces of teeth; and in lacking posterior crest of protocone of P^4. In place of posterior protocone crest, basal cingulum runs along posterior base of premolars forming a small but distinct posterointernal basin. Differs from other species of *Titanoides* in smaller size of M^3, relative to M^{1-2}; in possessing slightly more distinct paraconules and metaconules in conjunction with a smaller protocone on upper molars.

Discussion: This species occurs in the Gidley Quarry, which has also yielded specimens of *Pantolambda intermedius* and *Pantolambda cavirictus,* but there is no likelihood that the type specimen of *Titanoides simpsoni* could belong to either of these species. Of course, strict comparisons with *Pantolambda intermedius* are limited by the fact that the upper dentition of this form is unknown, except for one upper premolar, probably P^3 or P^4, from the same quarry and distinctly of pantolambdid type, which almost certainly belongs to this species. This premolar, AMNH No. 35724, resembles upper premolars of *Pantolambda bathmodon* and *Pantolambda cavirictus,* except that it is intermediate in size, but it is easily distinguishable from that of *Titanoides simpsoni.* The type lower jaw of *Pantolambda intermedius,* USNM 8384, from the Gidley Quarry, is so much like that of *Pantolambda bathmodon* that even if two kinds of pantodont upper premolars of this size range were not known from the Gidley Quarry, it would still be highly improbable that the lower dentition of *Pantolambda intermedius,* of strictly pantolambdid type, could belong with an upper dentition of titanoideid type such as of *Titanoides simpsoni.* The foregoing position as to the distinctness of these two forms would be weakened if it could be shown that the supposedly distinctive characteristics of the protocone crests of the upper P^{2-4} of *Titanoides* ever occur as a variant condition in pantolambdids. It should be pointed out that in all the pantolambdid upper dentitions examined, a total of over fourteen complete enough to exhibit this character, there is no suggestion of any variability in the direction of suppression of the crests of the upper P^{2-4} protocone. Nor does loss of these crests ever occur in individuals belonging to other pantodont species of North America, except in the Titanoideidae for which this character appears to be a diagnostic feature.

In spite of the similarities of *Titanoides simpsoni* with other species of *Titanoides,* the absence of the larger part of the dentition, the much smaller size, and the relatively small M^3 in particular render certain reference of this species to *Titanoides* more questionable than for the other members of the genus. Future discoveries may make generic separation of this form desirable, but it is hardly likely that they will contradict the titanoideid characters of the P^4-M^3.

Titanoides sp.

PU No. 14635

Horizon and locality: Melville formation, Crazy Mountain Field, Montana, about 2,000 feet above the Scarritt Quarry level, NE 1/4, SW 1/4, Sec. 19, R 14 E, T 5 N, Sweetgrass County, Montana.

Description: Four parts of a lower canine, two fragments of lower premolars or molars, part of a talonid of an M_3, and most of a right P^1.

Discussion: PU 14635 may be *Titanoides gidleyi,* but the specimen is too fragmentary to justify an assignment to a known species. The base of the canine preserved shows that the specimen is not *Titanoides primaevus.* Since these fragments are somewhat larger than in the type of *Titanoides gidleyi,* more nearly approaching teeth of *Titanoides majus* in size, they may pertain to the latter species, but assignment to this form would be premature at present.

The most important fact about this specimen is its high stratigraphic position in the Melville formation, about 2,000 feet above the Scarritt Quarry. Consequently, it is one of the youngest fossils from the Melville.

Titanoides ? sp.

CNHM No. P-26080

Horizon and locality: Plateau Valley fauna, DeBeque formation, one-half mile north of "Coffee Pot," Mesa County, Colorado.

Description and discussion: CNHM P-26080 from the DeBeque formation of Mesa County, Colorado, consists of both horizontal rami of the lower jaws, with the left ramus attached to the symphyseal region. All the teeth present are heavily damaged, except the left M_2 and M_3. These teeth are like those of *Titanoides* in all respects. The root of the left canine is preserved also; it is extremely small for a pantodont, proportionally smaller than the canine root of *Haplolambda quinni.* One might suppose that this is an immature individual of *Titanoides primaevus* in which the milk canine had not yet been replaced; but the fusion of the symphysis, known in no other specimen of *Titanoides,* would suggest that the individual was mature, or that fusion of the symphysis took place before the milk canine was replaced. In *Titanoides primaevus,* CNHM P-15520, however, in which the permanent canine has erupted, the mandibular symphysis is not yet fused.

In the present state of knowledge this specimen can only be classified as a titanoideid. It may or may not belong to the genus *Titanoides,* but the specimen is too incomplete to warrant making it a new genotype. The M_{2-3} characters show that it does not belong to any other family of pantodonts.

IV. THE PANTODONT SKELETON

INTRODUCTION

As a group the Pantodonta, of all the Paleocene Mammalia, are undoubtedly the best known skeletally, in spite of a comparatively small total number of individuals. Only *Ignatiolambda*, of the eight North American pantodont genera occurring in this epoch, lacks a reasonably complete skull and postcranial skeleton. It is, therefore, anomalous that apart from *Coryphodon* none of the Eocene and Oligocene members of this order have known postcranial remains. For this reason, and because of the abundance of published information on various Eocene species of *Coryphodon*, the following discussion of the pantodont skeleton will be limited largely to the North American Paleocene members of the order.

SKULL

Externally the pantodont cranium is in a general way like that of the other archaic mammals of the early Tertiary, particularly resembling those of such genera as *Deltatherium* and some of the primitive condylarths like *Ectoconus* and *Periptychus*. Moreover, this resemblance also extends to the brain cast for Edinger (1956 and personal communication) reports a correspondence between the endocranial casts of *Leptolambda* and *Periptychus*, at least in evolutionary grade. In overall appearance the pantodont skull agrees with other primitive mammals in its relatively small brain case, anterior position of the nasal opening, flaring zygomatic arches, pronounced saggital crest, and great post-orbital length.

The conformation of the premaxilla is reasonably distinctive in each of the four families of pantodonts. *Titanoides*, for instance, lacks an extension of the dorsal process of the premaxilla onto the lateral surface of the rostrum, and in this respect it agrees with *Haplolambda*, *Barylambda*, and probably also with *Leptolambda*. In the pantolambdids, on the other hand, there is a large facial extension or process of the premaxilla which is bounded posteriorly by an S-shaped suture and which extends dorsally to contact the anterior tip of the nasals. Among the Coryphodontidae this bone is much larger relative to the rest of the cranium, and provides a massive support for the upper incisors. In *Coryphodon*, also, the premaxilla has a facial extension, but in most individuals it does not reach the nasals, at least on the external surface of the skull. In all Paleocene pantodonts the premaxillaries join anteriorly in a pair of posteriorly directed wings which divide the anterior palatine foramina. Moreover, in all these forms, the premaxillae are slightly separated by an indentation between their anterior tips. In the Pantolambdidae all the incisors appear to be present and to increase in size from I¹ to I³ in all the species included in this family. Judging from the upper incisor alveoli present in the type of *Haplolambda quinni*, CNHM P-15542, this is

also the condition in some barylambdids, but in *Barylambda faberi*, the upper first incisor pair is usually missing.

The anterior nasal openings in *Barylambda* and *Haplolambda* are quite broad and flaring, a condition which suggests a large muzzle in these forms, but in *Pantolambda bathmodon* and *Caenolambda pattersoni*, the anterior nares are proportionately much smaller. The rounded contours and large size of the nasal opening of the skull of *Titanoides primaevus*, CNHM P-15520; suggest that in this species also there may have been a large and flexible snout. The proportionately very small upper incisors of *Titanoides* further corroborate this view, for if the upper lip and snout were enlarged they might take over part of the cropping function usually performed by the upper incisors—as among the Pecora. If the upper incisors of *Titanoides* were comparatively functionless, this might explain their small size. On the other hand, it is clear that in *Coryphodon* and the Asiatic members of the Coryphodontidae, the large incisors must have been fully functional as cropping teeth, and compared to *Pantolambda* they are relatively huge in proportion to the other teeth.

The ventral aspect of the muzzle differs little in the various Paleocene members of the Pantodonta. The exact positions of the posterior margins of the anterior palatine foramina are not known in any species. This portion of the skull is presumably preserved in one specimen of *Pantolambda bathmodon*, AMNH 16663, but because of the difficulty in getting the upper and lower teeth separated it has not been exposed. The posterior surface of the hard palate is similar in all of the various Torrejonian and Tiffanian pantodont species; and, as in some insectivorans, the posterior margin of the palatine between the anterior tips of the pterygoids is thickened. In the barylambdids the posterior palatine foramina are small and are scattered in the region adjacent to the lingual margins of the second upper molars, while in the most completely preserved palate of *Titanoides primaevus*, USNM 20029, there is a single posterior palatine foramen about one centimeter mesiad to the base of the upper second molar. The palatine bone is seen in the Chicago Museum specimen of this species Number P-15520, to have its anterior tip contacting the posterior margin of this foramen and to be separated from the maxilla by a suture running along to the back of the palate about .5 cm. internal to the base of the M²⁻³. In the most complete skull of *Pantolambda cavirictus*, AMNH 963, the region of the palatine is not preserved, but a referred specimen of the latter species, AMNH 17034, exhibits a single posterior foramen as in *Titanoides*.

The lateral aspect of the rostrum in the Torrejonian and Tiffanian pantodonts is approximately the same, except for the differences in the premaxilla mentioned above and that in *Titanoides* and *Barylambda* the maxilla is relatively deeper dorsoventrally.

Patterson (1935: fig. 1) indicates that in *Barylambda faberi* there is a facial extension of the lachrymal, which bears a distinct tubercle. The presence of this primitive feature in other Paleocene pantodonts has not been definitely observed although there is some suggestion of a facial extension of this bone in the skull of *Titanoides*, CNHM P-15520. The lachrymal foramen of this specimen, however, does not appear to be outside the orbit. In all the Paleocene members of this order there is a very large infraorbital foramen, as well as a relatively small orbit, bounded above by the anterior tips of the temporal ridges which are evident in all pantodont species but heaviest in *Caenolambda pattersoni* and *Barylambda faberi*. No distinct post-orbital process is known in any pantodont. The insertion of the anterior root of the zygomatic arch in the late Paleocene pantodonts is approximately above M^1 except in *Caenolambda pattersoni* where it is above P^2. The pantodont maxilla makes a long contact dorsally with the nasals which are typically expanded posteriorly in this order. Apparently, the maxilla in all pantodonts makes up part of the ventral border of the orbit below the facial extension of the lachrymal. The *pars orbitalis* of the maxilla consists of a horizontal plate which is triangular and has, in all Paleocene pantodonts, its greatest width at the posterior margin. Perhaps the greatest transverse diameter of the horizontal plate occurs in *Pantolambda cavirictus*. In *T. primaevus*, USNM 20029, the maxilla extends along the internal face of the zygomatic arch to a point opposite the posterior margin of the M^3.

The jugal in all species appears to possess two anterior wings; the upper being the longer and running along the posterior lower border of the orbit, while the ventral is more massive and is directed downward. In *Barylambda faberi* and *Haplolambda quinni* these two wings are less distinct. Patterson (1935: fig. 1) represents the anterior dorsal wing of the jugal of *B. faberi* as making up the entire ventral border of the orbit and as contacting the lachrymal directly. This region is very poorly preserved in Paleocene pantodont skulls, but at least in *Titanoides*, USNM 20029, the jugal appears to be separated from the lachrymal by a portion of the maxilla. The jugal in the pantodont skull extends for a considerable distance posteriorly beneath the zygomatic process of the squamosal. Furthermore, in *Barylambda*, in keeping with the greater massiveness of the whole arch, the entire posterior extension of the jugal is broad and in *Haplolambda quinni*, CNHM P-15542, it is slightly expanded at the posterior tip. On the other hand, in *Titanoides* and *Pantolambda*, where the arch is more slender and flaring, the jugal tapers posteriorly.

As Patterson has pointed out, the nasals of the Barylambdidae are distinctive. *Haplolambda*, *Leptolambda* and *Barylambda* all have nasals that are almost twice as broad in proportion to the rest of the skull as they are in *Pantolambda*, *Caenolambda*, and *Titanoides*.

Furthermore, in the barylambdids, particularly in *Barylambda faberi*, the posterior parts of the nasals are enormously extended laterally. In this respect also *Coryphodon* differs from the barylambdids in having slenderer nasals.

In all the Paleocene species the frontals are indented anteriorly by the broad posterior expansion of the nasals. As mentioned above, these bones bear well-developed temporal ridges and presumably contact anteriorly with the lachrymal and certainly with the maxilla. The ventral margin of the frontals is not well preserved in any Paleocene cranium I have seen, but at their posterior margin these bones have a V-shaped tip limited above by the parietals and below by part of the arcuate anterior margin of the squamosal. These features are most clearly observed in the Chicago Museum crania of *Barylambda faberi*. The frontals contact each other along the midline of the skull posterior to the point where the temporal ridges come together to form the anterior tip of the sagittal crest. From this point backwards the parietals make up the sagittal crest, which usually possesses a central groove. Moreover, the posterior dorsal tips of the frontals taper into a distinct V-shaped wedge between the anterior tips of the parietals; and in turn the parietals at least of *Titanoides primaevus*, *Pantolambda bathmodon*, and *Barylambda faberi*, are separated posteriorly by small interparietals. Their posterolateral tips, in *Barylambda faberi* and *Leptolambda schmidti*, extend along the middle third of the margin of the lambdoid crest. In all these Paleocene species the top of the sagittal crest has either a horizontal or slightly convex outline. This crest in *Titanoides primaevus* is quite high and narrow, though the extremely high and narrow sagittal crest of USNM 20029 shows an exaggeration of this feature due to distortion. The more typical structure of the crest can be seen in the better preserved specimen, CNHM P-15520. None of the North American Paleocene species of this order shows any suggestion of the extensive expansion of this crest into a broad flattened plate which is typical of the Coryphodontoidea. In the posterior region of the pantodont skull the parietals form part of the dorsal half of a concavity for the insertion of the huge temporal muscles; and they are extensively perforated, as is this same region of the squamosal, by numerous small nutrient foramina.

The squamosal is broadly expanded posterodorsally, particularly in *Titanoides*. It bears a strong zygomatic process which has on its ventral surface a considerable extension of the glenoid fossa. This surface, as well as the post-glenoid process, is not placed as in most recent mammals, at an angle perpendicular to the long axis of the skull. Instead, all the pantolambdoid species differ from *Coryphodon* in having the transverse axis of the glenoid fossa situated in a horizontal plane but at an oblique (instead of a right) angle to the long axis of the skull and with the external tip of the fossa pointed forward. In other words, the whole fossa has its main axis at an angle about thirty to forty degrees from a line transverse to the basicranium. As consequence of this arrangement the condyles of the lower jaws are necessarily also placed at an oblique angle. The post-glenoid processes in *Barylambda* and *Haplolambda* are high and about one-half shorter along their main axes than are those of *Titanoides*. A distinction of the latter genus in this respect is that the post-glenoid process is low, relatively broad, and long across the base, a feature which presumably correlates with the much wider gape of the mouth in *Titanoides*. *Coryphodon*, on the other hand, possesses a post-glenoid process which is much longer than it ever is in the Torrejonian and Tiffanian pantodonts. With regard to the width of this process, *Coryphodon* in most specimens resembles the Barylambdidae in that the length of the post-glenoid process transverse to the main axis of the skull is much shorter than in *Titanoides*. *Haplolambda* and *Barylambda* both possess a small accessory process at the posterointernal base of the glenoid process.

The zygomatic process of the squamosal in Paleocene pantodonts reaches its greatest depth at about half its length anterior to the base and from this point tapers rapidly. The anterior tip does not lie free above the jugal in *Titanoides* and *Haplolambda* but abuts against a posterodontal flange of the jugal. The contact of the squamosal at its internal forward extremity with the alisphenoid is not well preserved in Paleocene pantodont crania, nor is it clear how it joins the mastoid posteriorly. A process which may be the post-tympanic process of the squamosal is well developed in pantolambdoids and this, at its posteroexternal base bears a large foramen. In *Barylambda*, *Leptolambda*, and *Haplolambda* neither this process nor foramen is as large as in *Titanoides*. All these genera, however, with the possible exception of *Haplolambda*, show a low knob or accessory process at the posterointernal base of the supposed post-tympanic process. *Coryphodon*, on the other hand, has a much longer process here than in the pantolambdoids, and the accessory process is not present in the specimens examined in this paper.

Titanoides possesses a large condylar foramen just anterior to the base of each of the occipital condyles. In the skull of *Pantolambda cavirictus*, AMNH 963, this region is poorly preserved, but the condylar foramina seem to be similarly located. The basicrania of *Haplolambda*, *Barylambda*, and *Caenolambda* are also poorly preserved in the region adjacent to the occipital condyles; but the condylar foramina are apparently situated somewhat more anterior to the condyles than in *Titanoides*. In *Leptolambda schmidti*, PU 14996, this more anterior position of the condylar foramen is clearly seen.

The exact margins of the basioccipital are not precisely indicated in any of the Torrejonian and Tiffanian pandodont crania, except that in all these species the anterior margin of this bone can be distinguished by a

suture between it and the presphenoid which runs transverse to the long axis of the skull just behind a line connecting the posterior bases of the postglenoid processes. All the pantolambdoid pantodonts seem to differ from the Coryphodontoidea in having a relatively broad exposure of the anterointernal ventral face of the *os petrous*, which in these genera bears a large central prominence; but this region is usually not well preserved. In four specimens, however, three of *Titanoides primaevus*, PU 16490, USNM 20029, and CNHM P-15520, and one of *Leptolambda schmidti*, PU 14996, this area is preserved almost intact. The following comments are based principally on these four specimens.

In the specimen of *Leptolambda* the tympanohyal is present and is attached by a broad massive base. Its ventral surface is oval in outline and bears a central depression. In *Titanoides primaevus*, on the other hand, the tympanohyal is relatively much smaller and its base of attachment quite slender. Figure 15 illustrates the surprisingly well-preserved petrosal of the type specimen of *Titanoides primaevus* with suggested identifications of parts. The tympanohyal is restored from CNHM P-15520. Otherwise, except for minor retouching, the specimen is reproduced as it is. This seems to be the oldest well-preserved Tertiary petrosal. The basic impression which this bone gives is more suggestive of general ungulate affinities than of a relationship with the insectivore stock. There is also a resemblance in this bone to that of *Deltatherium*, and a much less striking but nevertheless suggestive resemblance to the petrosal of the aardvark. There is no evidence of an auditory bulla in any of the pantodonts in which this region of the skull is preserved. A tympanic is present in the type of *Leptolambda schmidti*, CNHM P-26075, but this is confined to the lateral part of the tympanic fossa and the remainder of the bulla was, no doubt, cartilaginous.

The posterior face of the occiput in all the Paleocene species has a semicircular dorsal outline. In *Pantolambda bathmodon*, *Haplolambda*, *Leptolambda*, and *Barylambda* as well, this surface of the occiput lies in a nearly vertical plane, but in contrast, *Titanoides*, and probably also *Pantolambda cavirictus* and *Caenolambda pattersoni*, have the dorsal tip of the occiput projecting considerably posterior to the occipital condyles.

PERMANENT DENTITION

The distinctions in the dentitions of the various Paleocene pantodonts have already been pointed out in some detail in the foregoing systematic revision, and certain other features of the teeth which can be functionally interpreted as adaptations are considered in the discussion of pantodont ecology in Section VI. The following remarks are, therefore, restricted principally to apparent evolutionary trends in dentitions of the various members of this order.

In *Coryphodon* and *Titanoides* the development of the protocone of P^2 is much more pronounced than in the Pantolambdidae and Barylambdidae. The latter two genera are also distinguished by the presence of a distinct protocone on P^1. The Barylambdidae and Pantolambdidae have a P^1 that is invariably one-rooted, but in *Titanoides* this tooth is always two-rooted. Some specimens of *Coryphodon* and the type of *Hypercoryphodon thomsoni* also exhibit a two-rooted upper P^1. Furthermore, among the coryphodontids and in *Titanoides* there is a much closer resemblance between the successive upper premolars than in the more primitive pantolambdids, *P. bathmodon*, and *P. cavirictus*.

Although there is a superficial likeness in P^1 of *Titanoides* and *Coryphodon*, this similarity does not indicate a close affinity nor does it seem possible that *Coryphodon* could be derived from any of the species of *Titanoides* as now understood. The distinction between the two groups is particularly evident in the structure of the more posterior upper premolars. P^{2-4} of *Coryphodon* have the typical pantodont paracone V and lingual protocone crescent with well-developed wings extending from the apex of the protocone along the anterior and posterior margins of the tooth and ending at the bases of the parastyle and metastyle respectively. *Titanoides*, on the other hand, lacks the posterior wing of the protocone entirely on P^{2-4}; instead the basal cingulum of the tooth rises on the lingual face of the protocone laterally, and it forms a distinct cingular shelf running along the base of the posterior half of the tooth.

In *T. primaevus* and *T. simpsoni*, the protocone of P^4 is shifted anterior to the transverse midline of the tooth and the size of the posterior cingular shelf, which is more strongly developed on P^{3-4}, tends to increase still more in these forms. The anterior wing from the apex of the premolar protocone is also disrupted in *Titanoides*, with the basal cingulum running along the anterior margin of the tooth. Nevertheless, there is usually some trace anteriorly of the primitive protocone wing above the continuous basal cingulum. Most probably these distinctions of the upper premolars of *Titanoides* represent a specialized rather than a primitive condition. Certainly, it is far more difficult to derive the upper premolars of *Coryphodon* from those of *Titanoides* than from the corresponding teeth of *Pantolambda bathmodon*, from which P^{2-4} of *Coryphodon* differ principally in the greater expansion of the anterior and posterior wings of the protocone, giving these teeth a more oval and less triangular outline.

Considering the relatively simple structure of the pantodont upper premolar, this distinctive development of the anterior and posterior basal cingula of the upper premolars of *Titanoides* represents a marked variation on the pantodont premolar pattern which had already arisen in Torrejonian times. This type of premolar distinguishes *Titanoides* from all other North American Pantodonta and is also clearly unlike the broken posterior protocone crest of the Mongolian *Haplolambda planicanina*. Considering the very minor advances in the pandodont dentition in other respects between the Torrejonion and Tiffanian stages, it does not seem likely that this specialization in *Titanoides* could have come about quickly.

MILK DENTITION

In figure 4 those elements of the milk dentition known in various species of pantodonts are illustrated. Of the Paleocene Pantodonta only the following forms have known milk teeth; first the type specimen of *Barylambda faberi*, CNHM P-14637, which includes the lower second and third milk premolars [30] and the upper second, third, and fourth milk premolars; second PU 16451 which is probably a lower DP_2 or DP_3 of *Titanoides*, third, PU 16459 a left milk DP^4 which is associated with fragments of M^1 of *Titanoides*. Various specimens of *Coryphodon* milk teeth are present in the Princeton collection and at the American Museum of Natural History. The best preserved of these is AMNH 15787 (of Eocene age). The restoration of the milk dentition included in figure 4 is based principally on AMNH 15787, with corrections for the broken parts derived from corresponding isolated teeth in the Princeton collections of Willwood Eocene age. Patterson (1934: 74) pointed out one of the most interesting features of the milk dentition of the type specimen of *Barylambda faberi;* that DP^4 is molariform and has a distinct paracone and metacone. This is also the case in *Coryphodon* and *Titanoides*. In *Barylambda* and *Titanoides* the upper fourth milk premolars, except for their smaller size, tend to resemble the M^1, while in *Coryphodon*, as Simpson (1929: 5) has suggested, the fourth upper milk premolar is transitional in its conformation between the upper molars of *Pantolambda* and the permanent molars of *Coryphodon* itself. This has been considered evidence of an affinity of *Coryphodon* with *Pantolambda*, and with the Torrejonian and Tiffanian species of pantodonts in general.

In *Barylambda faberi* the protocones of the permanent upper molars are shifted somewhat posteriorly relative to the rest of the tooth; while, on the other hand, the protocones of the upper permanent premolars are shifted anteriorly. In this specimen DP^4 resembles the permanent molars in having the protocone directed somewhat posteriorly.

The advanced condition of wear present in the milk molars of the type specimen of *Barylambda faberi* renders detailed comparisons with the milk dentition of *Coryphodon* difficult. Wear on the DP_3 of this specimen, for instance, has completely removed the crown of the talonid; so it is not clear whether this tooth was molariform or not, although the rectangular outline of the base of the talonid suggests that it may have been. This is not the case, however, in *Coryphodon*, AMNH 15787, where only the DP_4 has a molariform talonid.

---

[30] This tooth was not figured by Patterson (1933: fig. 1) because it had been dissociated from the specimen. Subsequently it was recovered.

The DP_2 in CNHM 14637 resembles its adult successor in *Barylambda* in having a very wide angle between the paraconid and metaconid crest in the trigonid. Furthermore, the milk tooth PU 16451, which is almost certainly a DP_2 of *Titanoides*, differs from the permanent tooth in this form only in its smaller size, and in having a groove on the internal face of the tooth between the paraconid and metaconid which is clearly more pronounced than in the permanent premolar of *Titanoides*. Lower milk premolars of *Coryphodon* are much like this tooth in having a similar pronounced indentation behind the paraconid. This character is also suggested in the much worn DP_2 of the type of *Barylambda faberi;* and, although the indentation behind the paraconid on the lingual margin of this tooth is not as great as that observed in corresponding milk teeth of *Coryphodon* and *Titanoides*, it is considerably more distinct than in the permanent P_2 of *Barylambda faberi*.

For whatever value comparisons between milk and permanent dentition may have, it should be pointed out that the DP^4 of *Coryphodon*, although possessing many of the distinctions peculiar only to that genus, is sightly more like a permanent M^1 and M^2 of pantolambdids than it is like the corresponding teeth in the Titanoi-

deidae and Barylambdidae. The upper milk second and third premolars have small protocones which are shifted somewhat posterior to the transverse axis of the tooth. They resemble very closely the P^2 of *Pantolambda cavirictus* in this respect, and again are less like the upper premolars of barylambdids and titanoideids. DP^1 of *Coryphodon* AMNH 15787 is two-rooted, judging from the alveolus. This is a condition which does not occur in the permanent P 1/1 in any barylambdid or pantolambdid. P^1 is, however, two-rooted in *Titanoides* as well as in *Coryphodon eocaenus* and occasionally in other species of *Coryphodon*. DP^4 in *Coryphodon* is molariform but does not exhibit any features which are departures from the general form of the M_1, except for smaller size. This tooth, like M_1 of *Coryphodon*, usually bears a cingulum on the anterior and posterior faces, has a reduced paraconid, large metaconid, and low and reduced *crista obliqua*. The talonid is also slightly broader than the trigonid in the specimens available.

DP^4 of *Titanoides*, PU 16459, resembles M^1 and M^2 of this genus principally in that the labial margin of the tooth is long anteroposteriorly in proportion to the transverse diameter, and in the proportions and position of the protoconule and metconule.

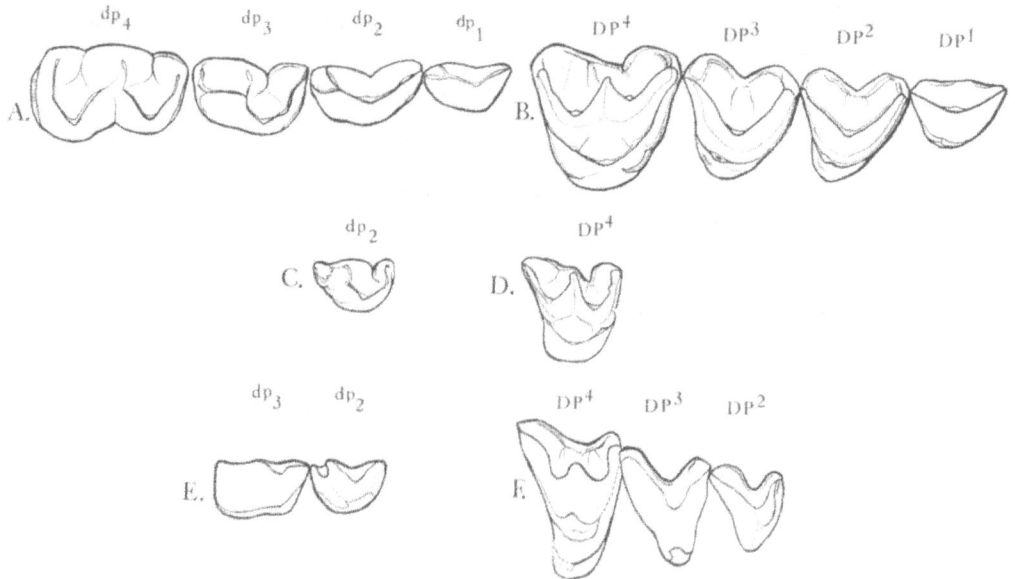

FIG. 4. Pantodont milk dentitions. × 1.

A. *Coryphodon* sp. based on AMNH 15797, lower right milk premolars. DP_1 added from other specimens in the AMNH and PU collections.
B. *Coryphodon* sp. based on AMNH 15787, upper right milk premolars. DP^1 added from specimens at Princeton.

C. *Titanoides ? gidleyi*, PU 16451, left DP_2(?) reversed.
D. *Titanoides ? gidleyi*, PU 16459, right DP^4.
E. *Barylambda faberi*, CNHM P-14637 (type), right DP_{2-3}.
F. *Barylambda faberi*, CNHM P-14637 (type), right DP^{2-4}.

MANDIBLE

In the foregoing systematic revision, emphasis has been put on the considerable variability in the conformation of the mandible of the Torrejonian and Tiffanian pantodonts. Practically the only diagnostic feature of the lower jaw shared by all these species is the position of the condyle at an oblique angle to the main axis of the horizontal ramus. The long axis of the condyle runs from the internal tip antero-externally.

The pantolambdid mandible is probably the most distinctive of the order. In this family the lower jaw possesses an antero-external flange at the base of the horizontal ramus of varying degrees of development which reaches its maximum size in *Caenolambda jepseni*. There is also an incipient development of this flange in *Titanoides primaevus;* but it does not project outward from the symphyseal region, when viewed from above, nearly as much as does the corresponding region in *Pantolambda cavirictus* and *Caenolambda*. This flange, however, is developed to about the same degree in *Pantolambda bathmodon* and *Titanoides primaevus,* so that the presence of this character is not entirely distinctive of the Pantolambdidae. No such flange is present in the Barylambdidae, but in *Leptolambda* there is a noticeable constriction in the mandibular ramus below the canine and premolars, which is probably analogous to the comparable concave region in the pantolambdid mandible above the flange. For comparison of this feature in the pantolambdid *Caenolambda jepseni* and the barylambdid *Leptolambda schmidti*, see figure 13.

In all of the Paleocene pantodonts the symphyseal region is relatively long, and the symphysis lies in a plane which is inclined anteriorly. The greatest length and anterior direction of the symphysis occurs in the Barylambdidae and particularly in *Barylambda faberi,* while the most vertical positioning of the symphysis in any Paleocene pantodont is that observed in *Titanoides primaevus*, CNHM P-15520. A further distinction of the titanoideids is that the mandibular symphysis does not fuse, as in the other three families, before the milk dentition is discarded, but remains open, if not permanently, at least until after the molars are fully erupted. No specimen certainly assignable to *Titanoides* is known which shows fusion of the symphysis. Presumably in connection with the presence of the jaw flanges beneath the lower canines in pantolambdids, the symphysis of the jaw is broad, strong, and massive. Barylambdids, on the other hand, possess a tapered symphyseal region, which in *Haplolambda quinni* is shallow and slender. With the exception of the latter species, and of *Haplolambda planicanina* as well, all pantodont species typically exhibit a constriction in the depth of the ramus posterior to the third lower molar. The position of the mental foramina in the Paleocene members of the order does not appear to be very consistent, and some individuals, for instance, *Leptolambda schmidti*, PU

14990, do not even agree in the location of these foramina on opposite sides of the mandible. In this specimen there is a double foramen below the left P_1 and a single opening below the right. The posterior foramina on the right side are almost confluent and are placed beneath the anterior root of the M_1. On the left side these posterior foramina are well over a centimeter apart. In pantolambdids and barylambdids alike there are usually one or more foramina below P_1, but in the pantolambdids they seldom occur behind this point. Barylambdids usually have mental foramina beneath M_1 as well. The one known complete mandible of *Titanoides primaevus,* that of CNHM P-15520, possesses a mental foramen beneath the posterior blade of the canine and beneath P_3.

Coronoid processes in well-preserved Paleocene pantodont mandibles are almost always directed backwards at an angle of about 115 degrees from a line drawn through the tooth row. These processes in the type of *Haplolambda quinni* are incorrectly attached to the rest of the mandible as restored, so that the backward slant is lost in this specimen. The correct backward tilt of the coronoid process in this genus is preserved in the specimen from the Polecat Bench formation, PU 16445. The strong backward tilt of the coronoid processes of *Barylambda faberi* CNHM P-14902, indicated in Patterson (1934: fig. 3), is much more pronounced than in the other specimens of this species and is surely due to distortion; Patterson later figured a mandible showing the correct inclination. Presumably Patterson's 1934 figure is the one Granger and Gregory (1934: 6) referred to when they compared the backward inclination of the ascending mandibular ramus of *Pantolambdodon* with *"Titanoides"* (*Barylambda*). In those forms with moderate or relatively small canines, such as the various barylambdid species, the coronoid process is relatively long; but in *Titanoides primaevus, Caenolambda jepseni,* and *Pantolambda cavirictus,* all of which species have elongate upper canines, the length of the coronoid process is much reduced—a feature which presumably facilitated wider opening of the mouth, as in the machairodonts.

Another characteristic of the mandible, which the latter three species share, is the reduction of the posterior extension of the jaw angle, also a feature to be correlated with the wider opening of the mouth. Conversely, the angle of the jaw has a considerable posterior extension in those pantodont species which have small or only moderately large canines; such as *Pantolambda bathmodon, Barylambda faberi, Haplolambda quinni,* and *Leptolambda schmidti*. In *Coryhodon,* the problem of opening the mouth wider to clear the large canines has been solved in another way than by the reduction of the posterior extension of the angle of the jaw. The zygomatic base has instead been shifted posteriorly toward the occipital condyles.

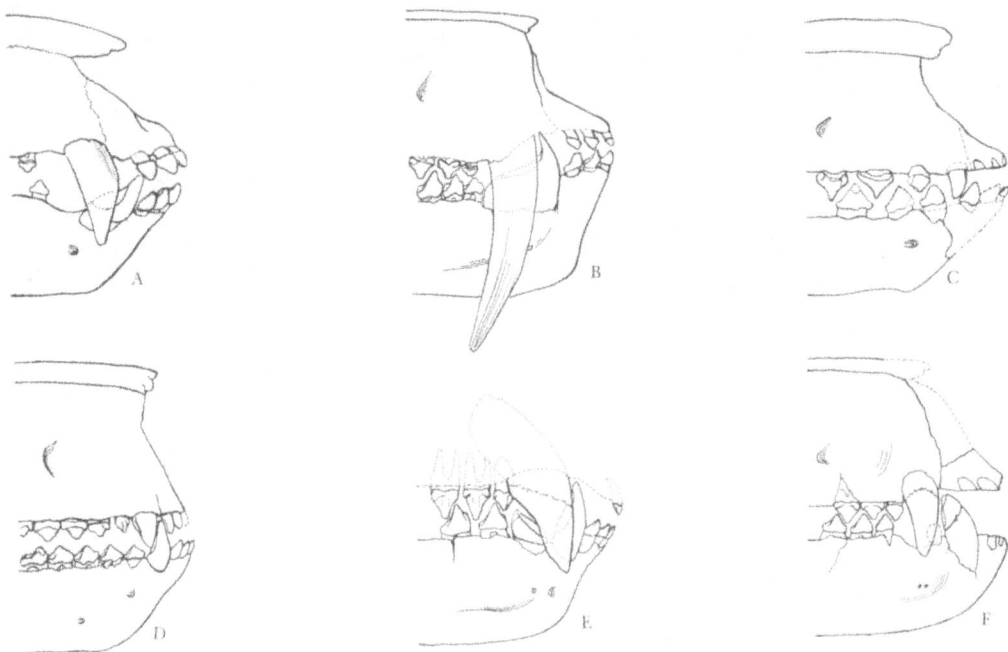

FIG. 5. Specializations of the pantodont anterior dentition. (Not to scale.)

A. *Coryphodon* sp., generalized diagram based on several specimens of Eocene age in the PU collections from the Bighorn Basin of Wyoming.

B. *Titanoides primaevus* Gidley. Upper dentition based on PU 16490 (type) and CNHM P-15520; lower dentition based on CNHM P-15520.

C. *Haplolambda quinni* Patterson. Drawn from the type specimen CNHM P-15542.

D. *Barylambda faberi* Patterson. Based on CNHM P-14902 and P-14944.

E. *Caenolambda jepseni* sp. nov. Drawn from upper and lower dentitions of type, PU 14863.

F. *Pantolambda cavirictus* Cope. Upper dentition and maxilla from U SNM 21327 (reversed); premaxilla from AMNH 16723. Lower dentition drawn from type, AMNH 3961, slightly reduced.

HYOID ARCH

Fragmentary bones believed to be the hyoid elements have been recovered with only one pantodont specimen, an individual of *Leptolambda schmidti*, PU 14680. If the identifications are correct, the bone considered to be the basihyal has a pointed median process which extends anteriorly. Posterior to the median process and between the articular region for the thyrohyals there is a pronounced concavity. Two much broken and distorted bones appear to be the parts of the thyrohyals proximal to the basihyal. They are not fused to the basihyal; but since this individual was not fully adult (judging from the incomplete closures of the epiphyses in the radius and ulna and other limb bones), it remains possible that in older individuals they would have anklyosed to the lateral extremities of the basihyal. A single flattened bone with a thin shaft and expanded at one end may be the distal portion of the stylohyal. As previously mentioned, the tympanohyal is known in *Titanoides* and the barylambdids. It

is a short bone with little or no neck, expanded, subcircular, and concave on its ventral face.

VERTEBRAE

The vertebral column in the Paleocene Pantodonta is remarkably well known, particularly among the Barylambdidae, and it is almost as completely known in the Titanoideidae and Pantolambdidae. For *Barylambda faberi* several virtually complete series of vertebrae have been recovered, together with the skeletons of this species, from the Plateau Valley beds of Mesa County, Colorado; and for *Leptolambda schmidti*, PU 14996, some cervical, thoracic, lumbar, and caudal vertebrae. Fragmentary vertebrae of the anterior half of the skeleton are preserved in the type specimen *Haplolambda quinni*, CNHM P-15542. The titanoideid spinal column is known principally from one specimen of *Titanoides primaevus*, CNHM P-15520, also from the Plateau Valley beds. This individual includes the first five cervicals, a number of broken and distorted thoracic and lumbar vertebrae, and three sacrals. There are also several isolated vertebrae of the titanoideid type from quarries in the Silver Coulee beds of the Polecat Bench formation of Big Horn County, Wyoming, as well as a few elements of the spinal column associated with other specimens of this family from the Plateau Valley, principally CNHM P-15547, P-15541, P-15523. In the Pantolambdidae the most complete spinal columns preserved belong to *Pantolambda bathmodon*, AMNH 16663, *Pantolambda cavirictus*, AMNH 2556, and the type specimen of *Caenolambda jepseni*, PU 14863. This last specimen includes seven cervicals, twelve thoracic, five lumbar, and two caudal vertebrae.

CERVICAL VERTEBRAE

In *Caenolambda,* the cervical vertebrae are particularly short and broad, and the neural spines are less robust than in *Titanoides* or *Coryphodon.* They appear to be almost as much reduced in this respect as in *Barylambda faberi.* In proportion to the lengths of the centra of the thoracic and lumbar vertebrae, the cervical central of *C. jepseni* are slightly shorter than in *Barylambda.*

The atlas of *Caenolambda* has the transverse processes broken off; but in the remainder of the specimen there is close confirmation in size and proportion to an atlas associated with a very fragmentary skeleton, AMNH 2556, presumed to belong to *Pantolambda cavirictus,* from the Torrejon beds of Escavada Arroyo, New Mexico. In the latter specimen surfaces for articulation with the condyles of the skull seem to be deeper and somewhat more recurved dorsally than in *Caenolambda.* The region of the neural spine of the atlas is not preserved in *Caenolambda,* but in the atlas of *Pantolambda cavirictus* there is only a slightly developed spinous process. The latter agrees in this respect with the atlas in *Leptolambda* and *Titanoides.* The atlas of *Caenolambda* also resembles that of *Pantolambda cavirictus,* AMNH 2556, except that the anteroposterior length of the latter specimen is somewhat greater, and the transverse process bearing the foramina for the vertebrarterial canal is much less compressed from front to back, than in *Caenolambda.* The spine of the axis in *Caenolambda jepseni* is broken along its dorsal margin, but its posterior extension does not appear to have been as long or heavy as in AMNH 2556, judging from the portion preserved. The median ridge at the base of the centra of AMNH 2556 is much more pronounced than in *Caenolambda,* and it is bifurcated posteriorly. The centrum of the third cervical in *Caenolambda jepseni* is longer than are the succeeding cervicals, all of which have bi-concave centra, the posterior concavity being the deeper. Transverse processes preserved in the second through sixth cervicals are all directed posteriorly, but are not as elongate or robust as, for instance, in *Ectoconus* or *Tillodon.* The sixth cervical vertebrae of *Caenolambda* exhibits the typical broad inferior lamellae of the transverse processes. Most of

the neural arch of the seventh cervical of this specimen is missing; but if it is properly pieced together (and this is not at all certain) it indicates that the transverse process of the seventh cervical is imperforate in *Caenolambda.*

The cervical vertebrae of the larger Torrejonian and Tiffanian pantodonts are distinctly divided into two types: the relatively small anteroposteriorly compressed vertebrae of the Barylambdidae; and the more massive cervicals of *Titanoides,* which have considerably longer centra, heavier laminae, and higher neural spines. In the latter group the transverse processes are also longer and directed more posteriorly than in *Caenolambda.* The cervicals of *Pantolambda* are about intermediate between those of the barylambdids and titanoideids in these respects; but in *Caenolambda* they are more like those of the barylambdids in their relative reduction, although the neural spines are higher. An isolated axis from Crocodile Tooth Quarry, in the Silver Coulee beds of the Polecat Bench formation, PU 16482, is very similar to that belonging to the Plateau Valley skeleton of *Titanoides primaevus,* CNHM P-15520; and these two vertebrae are distinctive in possessing a relatively great central length for a pantodont, the cervical centra being in PU 16482, over a half longer proportionately than in *Caenolambda jepseni* and *Barylambda.* Some specimens of *Coryphodon* also have a comparatively long centrum of the axis, but none are as long relative to width, as is the Crocodile Tooth Quarry specimen.

In absolute size, the atlas of *Coryphodon* is a third or more broader than that of the larger Tiffanian pantodont species, such as *Barylambda faberi* and *Leptolambda schmidti;* and in proportion to body size, it is greater still. This enlargement of the atlas and cervicals in *Coryphodon* is undoubtedly related to the proportionately larger head of this pantodont.

The prezygapophyses of the cervicals of *Titanoides primaevus* are slightly less widely expanded laterally than in other pantodonts, including *Coryphodon.*

DORSAL VERTEBRAE

A total of twelve dorsal vertebrae are preserved with the type specimen of *Caenolambda jepseni,* but this is presumably not the complete series. In *Coryphodon,* for instance, the number of dorsal vertebrae is seventeen, while in *Barylambda* it is fifteen. The total number of dorsals among the Pantodonta is known only in these two forms; but a few elements of this series have been recovered with skeletons of *P. bathmodon, Titanoides,* and *Leptolambda* as well.

In the anterior dorsals of *Caenolambda jepseni* the spines are long and are strongly directed backwards; but in the last three of the series, they are shorter, elongated from front to back and slightly expanded at the tips. In the last two members of the series preserved in *C. jepseni,* PU 14863, they are placed vertically. The backward direction of the spines ends with the eighth dorsal of PU 14863, and it is presumably at this point that some vertebrae are missing. On the most anterior dorsals of *Caenolambda* the facets for the articulation of the heads of the ribs face almost directly anteriorly and posteriorly on the anterolateral and posterolateral margins of the centrum. The transverse processes increase in size posteriorly, and after the second or third dorsal, are relatively high, as in *Barylambda*, and are directed backward and slightly inward at their tips. The neural spines of the dorsals of *Barylambda faberi*, CNHM P-14945, are comparatively massive. In *Coryphodon* the height and thickness of these spines vary considerably, but in the spinal column of *Coryphodon* figured by Osborn (1898b: fig. 23) they are relatively large. The postzygapophyses of the dorsals of *C. jepseni* are larger than the prezygapophyses and are somewhat concave. In the most posterior members of the dorsal series of this species the anapophyses and metapophyses increase in size toward the back. The dorsal metapophyses in *Caenolambda* form rounded knobs but are not expanded at their tips as in *Trogosus*, while the anapophyses are elongate posteriorly and pointed. In proportion to the length of the lumbars, the average length of the dorsal centra of *Caenolambda* is slightly less than in *Barylambda*. Patterson (1939a: 357) remarks however, of "*Sparactolambda*" [= *Titanoides*] that the ". . . dorsals and lumbars agree with those of *Pantolambda* and *Coryphodon* in having much longer centra than those of *Barylambda*." It is clear that, at least in the best-known pantolambdid spinal column, the relative proportions of the lengths of the dorsal and lumbar centra are much closer to those of the Barylambdidae than to the other pantodont families. However, all barylambdid centra are relatively high in proportion to length. None of the specimens of *Pantolambda bathmodon* at the American Museum includes a very complete dorsal series; and of the best preserved specimen Matthew (1937: 176) remarks:

> The three last dorsals are much as in *Ectoconus*, except that the centra are rounded beneath without median ridge and the anterior zygapophyseal process is less prominent and is compressed into an oblique plate . . . .

In this regard the posterior dorsals of the specimen here referred to *P. cavirictus*, AMNH 2455, are somewhat intermediate between *P. bathmodon* and *Caenolambda* in that there is only a slight suggestion of a median ridge. This ridge in *Caenolambda* is strongly developed. In both AMNH 2455 and *Caenolambda*, PU 14863, the lateral faces of both dorsal and lumbar vertebrae are deeply excavated. Both specimens also agree in having the anterior and posterior faces of the dorsal vertebrae concave. In *Caenolambda* the posterior concavity is noticeably deeper, but in AMNH 2455 the reverse is true. In both of these pantodonts the facets for the rib tubercles face anterolaterally.

The few known dorsal vertebrae of *Titanoides* are not well preserved, but they appear to resemble those of *Caenolambda* a little more closely than they do those

of *Barylambda*, particularly in having shorter spines. As mentioned already, however, they have proportionately longer centra.

LUMBAR VERTEBRAE

The specimen of *Pantolambda bathmodon*, AMNH 16663, preserves five lumbar vertebrae; and this is also the number of lumbars associated with the type of *Caenolambda jepseni*, as well as with specimens of *Barylambda faberi*. Osborn (1898a: 85) gives the number of lumbars in *Coryphodon ?testis*, AMNH 2829, as five; but Patterson (1939b: 102) reports only four lumbars in a specimen of *Coryphodon ?elephantopus*. It would appear that the primitive number of lumbar vertebrae in this order must have been five.

The relative positions of the five lumbars preserved in *P. bathmodon*, AMNH 16663, are not certainly known; and it is evident the series is not complete in the provisionally referred specimen of *P. cavirictus*, AMNH 2455, notwithstanding Wortman's remarks to the contrary. It is, therefore, fortunate that the type of *Caenolambda jepseni* includes all five members of the series. As in *Barylambda*, the transverse processes of the lumbar vertebrae of *Caenolambda* increase in size posteriorly. These processes in the posterior three members of the series (in both forms) are quite flattened and long, and their extremities are directed slightly anteriorly. In the first and second lumbars of *Caenolambda jepseni* anapophyses are present arising on the posterolateral bases of the transverse processes, but they are not distinct in the remainder of the series. The lumbar neural spines of the type specimen, PU 14863, are not as high as those of the barylambdids; but in proportion to the length of the centra of the rest of the vertebral column they are about as long as in the barylambdids. The depth of the centra is, however, much less in the lumbars of *Caenolambda jepseni*. The neural spines of these vertebrae in *C. jepseni* have their axes placed vertically as in AMNH 2455 (*Pantolambda ?cavirictus*) and in *Barylambda faberi*; but in *Caenolambda* their apices are apparently more broadly expanded than in other pantodonts and are triangular in outline when seen from above, with the point directed forward. A similar condition is present in the lumbar neural spines of the Polecat Bench *Coryphodon* PU 14685; but in this individual, as was noted by Osborn (1898a) for Eocene specimens of this pantodont, the lumbar prezygapophyses are set in an almost vertical plane. In *Caenolambda*, on the other hand, the zygapophyses are more broadly expanded so that the postzygapophyses face outward and much more vertically than in *Coryphodon*. In the latter form the neural spines of the lumbar series are directed even more posteriorly than in pantolambdids or barylambdids. In PU 14685 the preserved lumbar transverse processes are relatively thick dorsoventrally and near their bases are constricted from front to back so that they are approximately circular in cross section. At their tips, however, they are expanded. In contrast to this, these processes in *Caenolambda* are broad and flat at their bases and apparently taper toward their tips. *Barylambda faberi*, AMNH 32511, on the other hand, possess a distinct neck or constriction near the base of the lumbar transverse process. This is particularly evident on the fifth lumbar of the American Museum specimen.

The lumbars of *T. primaevus* are rather like those of *Caenolambda* in conformation and in proportion to their total height much longer anteroposteriorly through the centrum than those of *Barylambda*.

SACRUM

There are three sacral vertebrae associated with the pelvis of *Pantolambda bathmodon*, AMNH 16663. This also appears to be the usual number of sacrals in *Barylambda faberi*, and it may be that three is the typical number of sacrals in this order.

The sacrals of *Pantolambda bathmodon* are distinctive in that they lack the fusion of their neural spines which occurs in some of the later pantodonts. Moreover, a fragment of the last sacral of the provisionally referred specimen of *Pantolambda cavirictus*, AMNH 2455, indicates that at least the spine of the most posterior sacral in this form, was not fused with those anterior to it. Nevertheless, the number of sacrals cannot be determined in this specimen, nor is their number known for certain in *Titanoides primaevus*. The Paleocene pelvis of *Coryphodon* illustrated in Figure 16 has three sacral vertebrae and a pseudosacrum posterior to it which includes at least four of the anterior caudals. Characteristically *Barylambda faberi* has three sacral vertebrae, but in one specimen, CNHM P-14944, a fourth vertebra is fused to the sacrum. The number of sacral vertebrae in *Leptolambda schmidti* is also three; and, as in *Barylambda*, two adjacent sacrals of the immature specimen of *Leptolambda*, PU 14996, have their neural spines fused. In *Barylambda faberi* the two anterior sacral spines are usually fused but are separate from the neural spine of the last sacral vertebra. In all the Paleocene pantodonts, with the exception of *Coryphodon*, PU 14685, the postzygapophyses for articulation with the anterior caudal are distinct. These species also agree in that the transverse processes of the sacrals are expanded at their lateral margins and fused together from front to back, forming a rigid sacral mass for articulation with the ilia. Apparently co-ossification of the sacrum with the ilia is complete in *Coryphodon* and *Barylambda faberi*, even in young adults; but in the pelves of *Titanoides primaevus* and *Pantolambda cavirictus* the ilia do not appear to have ankylosed completely with the sacrum.

CAUDAL VERTEBRAE

Two anterior caudal vertebrae associated with the type of *Caenolambda jepseni*, PU 14683, resemble very closely those associated with the pelvis provisionally referred here to *Pantolambda cavirictus*, AMNH 2455. The centra of these vertebrae are short compared to those of some Creodonta and Condylarthra, but not nearly as short proportionately as the anterior caudals of *Leptolambda*, PU 14996, and of *Barylambda*. The two caudals recovered with *Caenolambda*, PU 14863, have cylindrical, bi-concave centra, complete neural arches and strong imperforate transverse processes. That these two caudals belong in the first three or four of the series is suggested by the fact that from about the fifth caudal on back in *Barylambda*, *Leptolambda* (and presumably other pantodonts) for the next eight to ten vertebrae the transverse processes are perforated, but there is no perforation of the process anterior to the fifth or sixth. The two caudals of *Caenolambda* preserved are large and lack perforations of the transverse processes, consequently suggesting an anterior position for them. As Wortman (1897: 84) mentioned with reference to the specimen AMNH 2455 (here referred to *Pantolambda ?cavirictus*) chevron bones are present on the anterior caudals. Although the two caudals of *Caenolambda* found are from this region of the tail, no chevron bones were recovered with this specimen. It is interesting to note the presence of chevrons in *Pantolambda*, as they have formerly been considered diagnostic of the Barylambdidae alone (see Patterson 1939a: 362). It is therefore curious that no chevrons have been recovered with any of the specimens of the smaller species of *Pantolambda*. The transverse processes of the anterior caudals of both *Caenolambda jepseni* and *Pantolambda cavirictus* have their tips directed downward and posteriorly; and, in proportion to the vertebrae as a whole, they are much larger than in *Barylambda* or in *Leptolambda*. In the latter two genera the laminae of the neural arches and the neural spines are higher than they are in *Caenolambda* and *Pantolambda cavirictus*. In the latter two pantolambdids the neural spines apparently die out after the second or third caudal but in *Barylambda* approximately the first ten caudals have well-developed neural spines. It is certain that in this genus the chevrons as well are more numerous. This vertically expanded and laterally compressed tail in *Barylambda* is clearly a highly distinctive specialization for this order.

In contrast to the relatively huge tails of the Barylambdidae and Pantolambdidae, which are nevertheless quite distinct from each other, *Titanoides* resembles *Coryphodon* in having a much-reduced tail, as has been emphasized by Patterson (1939a: 367). A single isolated caudal from Crocodile Tooth Quarry, PU 16487, resembles closely those associated with specimens of *Titanoides* at the Chicago Natural History Museum, but is asymmetric, the right transverse process only being perforated. This caudal and others belonging to *T. primaevus* at the Chicago Museum are all strikingly like those of *Coryphodon*.

RIBS

For most of the Paleocene species of pantodonts at least some of the ribs are known, but, with the exception of some individuals of *Barylambda faberi,* few pantodont specimens are complete enough in this respect for really detailed comparisons. Parts of about twelve ribs were recovered with the type of *Caenolambda jepseni,* but none of them are complete. However, the most anterior of these are flattened along the shaft, expanded distally, and the tubercles are distinctly set off from the heads. Fragments, presumably belonging to more posterior ribs, are oval in cross section. The ribs associated with the mounted skeleton of *Pantolambda bathmodon* at the American Museum of Natural History are not really complete enough to make comparisons possible. This is also true of the ribs recovered with the specimen of *Titanoides primaevus* from the Plateau Valley CNHM P-15520. The ribs seen in the mounted skeletons of *Barylambda faberi* are shorter compared to the body as a whole than in *Coryphodon;* but, as Patterson (1939*b*: 104) indicates, there is apparently some variability in the length of the ribs within the latter genus.

CLAVICLE

The clavicle is known in *Coryphodon* as well as in all of the earlier genera of pantodonts, except *Ignatiolambda.* As might be expected among the earlier Torrejonian and Tiffanian forms, the clavicle is relatively large and well developed compared to the condition of this bone seen in *Coryphodon.* Probably the most distinctive form of the clavicle among the Pantodonta is that of *Titanoides,* which has an extremely flattened shaft. In the type of *Caenolambda jepseni,* on the other hand, this bone has a much thinner shaft that is oval in cross section and larger at the sternal end. In *Leptolambda schmidti,* PU 14996, it is intermediate in size between that of *Caenolambda* and of *Barylambda.* All barylambdid clavicles resemble that of *Barylambda* figured in Patterson (1934: 81). Concerning the clavicle of *Pantolambda bathmodon,* Matthew (1937: 177) remarks:

> The clavicle is a long and heavy bone, moderately curved, of round-oval, cross section somewhat widened at the sternal end, widening into an oblique-ending expansion at the acromal end. The bone is quite as heavy as in the beaver or in man, and has none of the sigmoid curvature of the primate scapula. It does not differ materially from the clavicle of *Periptychus* or *Ectoconus.*

In *Barylambda, Leptolambda,* and *Caenolambda,* but not in *Titanoides,* the clavicle is expanded into a round knob-shaped process at the acromial end, which is seldom well preserved. Its articular relationship with the tip of the acromion process of the scapula is uncertain, but the tip of the acromion tapers to a much smaller apex than does the lateral end of the clavicle. Presumably these two bones had a cartilagenous attachment.

SCAPULA

Matthew's description of the scapula of *Pantolambda bathmodon* (1937: 177) emphasizes its primitive character. The specimen which he described, AMNH 16663, demonstrates that the restoration of the scapula of *Pantolambda* published by Osborn (1898*b*) and based on that of *Coryphodon* was in error.

The scapula of *Caenolambda jepseni,* PU 14863, is extremely well preserved, except for the tip of the acromion and part of the suprascapular border (see fig. 6). It agrees in general proportion with the scapula of *Pantolambda bathmodon,* AMNH 16663, in those parts which can be compared, except that the region of the metacromion in *Caenolambda* is somewhat more produced posteriorly; and there is a distinct tuberosity directly anteriorly and placed slightly above the middle of the spine. Both of these specimens agree in that the suprascapular notch is not as deeply indented as it is in the Barylambdidae. This is probably also true of the scapula of *Titanoides primaevus,* CNHM P-15520. The scapula of the type of *Caenolambda jepseni* has a distinctive thickening at both the coracovertebral and the glenovertebral angles which is much more pronounced than in the scapulae of *Barylambda faberi,* CNHM P-14902, and in *Leptolambda schmidti,* PU 14990. Furthermore, the glenovertebral angle of the scapula of *Caenolambda* is not as produced posteriorly as in *Barylambda faberi,* nor are the supraspinous fossae and infraspinous fossae as broad as in *Leptolambda* and *Barylambda.* The coracoid process of the scapula of *Caenolambda* has been somewhat damaged; but it does not appear to have been as massive or as long as in *Leptolambda,* PU 14990, or in *Barylambda,* and in this respect agrees with the poorly preserved scapula belonging to the skeleton of *Pantolambda bathmodon,* AMNH 16663. The scapula of *Titanoides primaevus* is known only partially; but the preserved portions of this bone belonging to the Plateau Valley skeleton, CNHM P-15520, shows that the coracoid process in *Titanoides* differs from the other Paleocene Pantodonta in its comparatively huge size and thickness.

The scapula of *Leptolambda schmidti* is best known from a complete specimen in the Princeton collections, No. 14990 (see fig. 6) which is much more like the somewhat larger scapula of *Barylambda faberi* figured in Patterson (1934: 82) than that of *Caenolambda jepseni.* This scapula of *Leptolambda,* PU 14990, is about twenty per cent shorter in overall length, parallel to the spine, than is the scapula of *Barylambda,* CNHM P-14902, and compared to its total length along the spine is only two-thirds as broad as in the latter specimen. In keeping with its generally smaller size, the acromion and coracoid process of the scapula of *Leptolambda* are slightly less massive than in scapulae of *Barylambda faberi,* and the scapular spine appears to differ from that of the latter species in its greater narrowness at the center. Furthermore, the scapular spine of *Leptolambda* in the most dorsal third of its length slightly overhangs the infraspinous fossa. The scapula of *Leptolambda*, PU 14990, at Princeton appears not to differ in any of the above respects from the two less well-preserved scapulae of *Leptolambda schmidti* at the Chicago Natural History Museum. These belong to two partial skeletons of this species, CNHM P-26077 and P-26098, recovered from the Plateau Valley beds. The scapula of CNHM P-26098, however, is over fifteen per cent smaller than those of PU 14990 and CNHM P-26077.

In discussing the scapula of the pantolambdoids, the marked distinctions in the scapula of all these forms from the conformation of this bone in *Coryphodon* should be stressed. Patterson (1939b: 105) has suggested, following Gregory, that the distinctiveness of the scapula of *Coryphodon* is in large part due to the need for a large base for attachment of the dorso-scapular ligament over the pointed apex of this bone (situated in *Coryphodon* in the center of the suprascapular border) in order to support the comparatively large head and neck. This supposition seems to be confirmed by discovery of the scapulae of the relatively small-headed pantodonts mentioned above—these lack the pointed apex and proportionately much narrower supraspinous and infraspinous fossae characteristic of *Coryphodon*.

The triangular apex of the scapula in *Coryphodon* is almost certainly an ossification of the suprascapular cartilage, as is evidenced by the fact that in well-preserved scapulae of this genus the pointed apex is separated from the body of the scapula by an incompletely fused suture. Such a suprascapular cartilage may also have been present in the Pantolambdoidea, but it is not known to have been ossified. Most scapulae of *Coryphodon* also differ from those of Torrejonian and Tiffanian pantodont species in that the subscapular fossa is markedly concave; so that when the scapula is viewed from the anterior side, the base of the spine is seen to be curved inward at the tips.

HUMERUS

Several virtually complete humeri of *Barylambda faberi* are preserved in the collections of the Chicago Natural History Museum, the best specimens available for study being those belonging to numbers P-14902 and P-26110-1. Reasonably well-preserved humeri of *Leptolambda* were recovered with PU numbers 14996, 14990, and 14680, as well as with CNHM P-26077. A virtually complete humerus of *Titanoides primaevus* associated with CNHM P-15520 is known, as well as partial humeri associated with other specimens of this species. The humerus of *Caenolambda* is not known, and of the Torrejonian pantodonts the only well-preserved humeri belong to *Pantolambda bathmodon*, although a distal portion of the humerus of *Pantolambda cavirictus* is associated with AMNH 964, and there is a distal third of the humerus, PU 16488,

presumably of this species from the vicinity of the Gidley Quarry. The proximal portion of the humerus associated with AMNH 2556 is not particularly like these but the differences in it may be due to distortion and erosion of the surface.

All these humeri are remarkably similar considering the rather radical adaptive distinctions present in the pantodont forefoot, differing principally in the robustness of the shaft and the height of the deltoid and supinator crests. The head of the humerus in all these forms is massive and is directed backwards. In all Paleocene pantodonts the deltoid and supinator ridges are high and an entepicondylar foramen is present, while the internal tuberosity is closely placed to the head and smoothly rounded. Further, the greater tuberosity is separated by a bicipital groove of moderate depth. The deltoid ridge in pantolambdoids appears to end at the base of the greater tuberosity, but in *Barylambda* and *Leptolambda* it is remarkably thickened from front to back at about the middle of the shaft. In these two forms also the width across the distal condyles is much greater in proportion to the length of the whole bone than it is in *Pantolambda*, and this is particularly so in *Titanoides*. In one specimen of *Titanoides*, CNHM P-15547, the entepicondylar foramen is relatively much reduced. This may or may not be a significant approach to *Coryphodon* in which, presumably because of the acquisition of subgraviportal structure, the entepicondylar foramen is almost never present.

Concerning the humerus in *Coryphodon*, Patterson (1939b: 106) remarks:

> The greater tuberosity is very prominent but does not extend dorsally above the level of the head; the lesser tuberosity is lateral in position. The bicipital groove is wide and shallow. The proximal end is by no means as thick anteroposteriorly as that of *Barylambda*, the greatest diameter being antero-internal-postero-external. The prominent, laterally inclined deltoid crest extends down from the great trochanter to the distal fourth of the bone. It resembles that of *Pantolambda*, being much more slender than in *Barylambda*. There is no latissimus tubercle. The supinator crest is similar to that of *Pantolambda*, and, as in that genus, a rounded concavity is present on the anterior face above the trochlea. The entocondyle is the most interesting part of the bone. It is larger and more prominent than that of any specimen of *Coryphodon* hitherto figured, and bears a small entepicondylar foramen—a structure never before recorded for the genus. This foramen occurs in the Paleocene genera, and it is interesting, and taxonomically significant, to find that it continued over into the lower Eocene.

RADIUS

Complete radii are associated with several specimens of *Barylambda*, *Titanoides primaevus*, the type of *Haplolambda quinni* in the collections of the Chicago Natural History Museum, and with *Leptolambda schmidti*, PU 14996, 14680, and 14879, from the Bighorn Basin, Wyoming. A complete radius is associated with the skeleton of *Pantolambda bathmodon*, AMNH 16663,

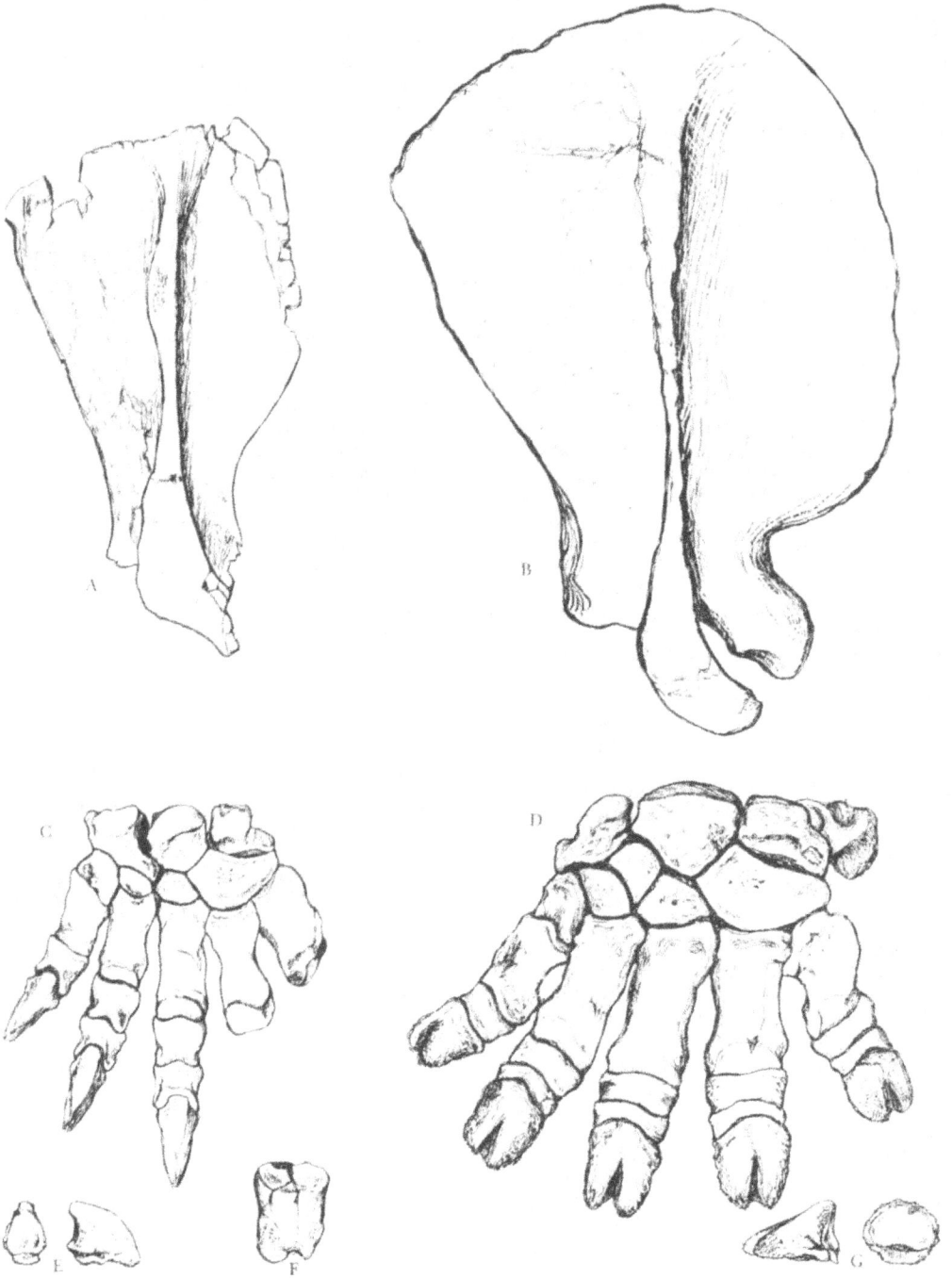

and an almost entire radius, probably of *Pantolambda cavirictus*, from the Crazy Mountain field, PU 13759, has been figured by Douglass (1902: pl. XXIV); but this bone is not known in *Caenolambda* or *Ignatiolambda*. A referred specimen of *P. cavirictus*, AMNH 16042, includes a complete radius which is similar in size and conformation to that of the type of *Haplolambda quinni*.

These radii, as well as that of *Coryphodon*, agree in lacking a pronounced constriction at the base of the head; nor do they exhibit a very prominent tuberosity for the biceps muscle on the posterior proximal surface of the shaft. They further agree in showing a gradual increase in the diameter of the shaft distally. The articular surfaces for the humerus and ulna are remarkably similar in all these forms, and both these surfaces extend completely across the head of the radius.

In *Pantolambda bathmodon*, AMNH 16663, the shaft of the radius (in proportion to the width of the whole bone) is slender, and seen from the medial side describes a slight sigmoid curve not present in any of the succeeding forms, or in the referred radius of *Pantolambda cavirictus*, PU 13759. This curve of the radius of *P. bathmodon* is emphasized by an extremely high and elongate crest on the anterodistal face of the bone. This distinct crest is also present in some specimens of *Leptolambda*, particularly in PU 14996. It is not nearly as distinct in the radius of *Leptolambda*, PU 14680. Comparisons with *Barylambda* and *Haplolambda* indicate that in these forms this crest is much less distinct, and it is apparently never present in *Titanoides* or in *Coryphodon*. The latter two types also agree in having a much shorter radius than in *Leptolambda* or *Barylambda*.

In all of the Tiffanian pantodonts, the distal extremity of the radius is broad; and the styloid process is heavy and considerably extended anteromedially. The articular surface for the lunar in *Leptolambda* is an oval concavity. In *Titanoides primaevus* and some specimens of *Coryphodon* this depression is much more circular. Radii of *Leptolambda* and *Barylambda* have on their anteromedial surfaces a wide depression for

the radial extensors and on the anteroexternal face an equally distinct groove for the digital extensors. These two depressions are separated in *Leptolambda schmidti*, PU 14996, by a prominent tubercle. As might be expected in the presumably much more rigid forefoot of *Coryphodon*, these grooves for the extensor muscles and the central tubercle are not at all distinct; see for example the radius of *Coryphodon* belonging to AMNH 2829.

ULNA

Among the ulnae of *Pantolambda bathmodon* preserved in the collections from the Torrejon beds at the American Museum of Natural History, the best preserved is that of AMNH 16663. An ulna figured by Douglass in 1902, PU 13759, from Bear Butte, Montana, is of the size and conformation expected for *Pantolambda cavirictus*. The ulnae of several specimens each of *Titanoides primaevus*, *Barylambda faberi*, *Haplolambda quinni*, and *Leptolambda schmidti* are included in the collections of the Chicago Natural History Museum from the DeBeque formation of Mesa County, Colorado; and three ulnae of *Leptolambda* from Silver Coulee beds of the Polecat Bench formation, belonging to PU Numbers 14680, 14996, and 14879, are included in the Paleocene collections of Princeton University.

The shaft of the ulna in *Pantolambda bathmodon* and in *Pantolambda cavirictus*, PU 13759, is broad and flat, particularly below the sigmoid cavity. Both the transverse compression and breadth of the shaft are comparatively greater in these Torrejonian pantodonts than in the Tiffanian forms. The olecranon process in *Pantolambda bathmodon* is also broader relative to the length of the shaft, than in the later forms. The sigmoid notch in *Pantolambda*, however, is apparently not as broad in proportion to the length of the shaft as in *Leptolambda* and the other Tiffanian species.

In all the Paleocene Pantodonta this bone exhibits a prominent interosseous crest running along the inner face of the shaft. In *Leptolambda* and *Barylambda* and to a lesser extent in *Titanoides primaevus*, the greater sigmoid notch is broadly expanded lateromedially and the coronoid process is large and produced anteromedially. An anterolateral process also projects from the base of the sigmoid notch, and the anconeal process is high and describes a pronounced S-shaped curve. At its dorsal margin in *Leptolambda* this process gives rise to a ridge running dorsointernally across the anterior face of the narrow olecranon process. In this genus also the smaller sigmoid cavity presents a continuous surface for the articulation of the head of the radius, but the distal facet for the articulation of the radius is not well preserved in any of the Paleocene specimens I have seen. The Tiffanian pantodonts all agree in that the styloid process is set off by a constriction situated principally on the internal face of the distal extremity. This extremity is rotated inward,

FIG. 6. Elements of the pantodont forelimb. × 1/3.

A. *Caenolambda jepseni* sp. nov. PU 14863, right scapula.

B. *Leptolambda schmidti* Patterson and Simons. PU 14990, right scapula.

C. *Titanoides primaevus* Gidley. CNHM P-15520, left manus with unguals restored from those associated with CNHM P-15547 and CNHM PM-240.

D. *Leptolambda schmidti* Patterson and Simons. PU 14996, left manus; scaphoid based on fragment from right manus, and on scaphoid of PU 14879; pisiform restored from an isolated specimen PU 16485.

E. *Titanoides primaevus* Gidley. CNHM PM-240, proximal and lateral aspects of claw of first digit.

F. Pantodont indet., PU 16483, proximal phalange resembling that of *Titanoides*, but much larger.

G. *Leptolambda schmidti* Patterson and Simons. PU 14996, proximal and lateral aspects of ungual.

and bears a continuous, rounded surface for articulation with the cuneiform.

The ulnae of *Leptolambda* in the Princeton collections differ from those of *Barylambda faberi* in having a shaft which is less broad anteroposteriorly, compared to the breadth of the olecranon process, and which is slightly retroflexed. *Barylambda*, on the other hand, has a straight ulnar shaft; and the greater sigmoid cavity appears to be placed somewhat higher, so that the olecranon process is comparatively shorter than in *Leptolambda*. The ulna in *Titanoides primaevus* differs from this bone in the Barylambdidae principally in its greater stockiness and in possessing heavier ridges on the lateral surface below the greater sigmoid cavity.

MANUS

The pantodont manus is not as well known as is the pes for there is no representation of the forefoot of *Caenolambda*, and in *Pantolambda cavirictus* only one element of the manus has been recovered, a cuneiform associated with AMNH 16042, provisionally referred to this species. Cope, Osborn and Matthew, have previously figured and described the manus of *Pantolambda bathmodon*, which is well represented in AMNH Numbers 2546, and 16663. However, the centrale of this species has apparently never been recovered. Its existence, nevertheless, is clearly indicated by the conformation of the surrounding bones.

The situation with regard to knowledge of the barylambdid manus is much more satisfactory, for essentially complete forefeet are known in most species belonging to this family. Furthermore, most of the left forefoot of *Titanoides primaevus* was recovered with the Chicago Museum skeleton of the species and this specimen is further supplemented by elements of the manus associated with three specimens at the Chicago Museum, P-15547, P-15541, and PM-240. Moreover, the Plateau Valley dentitions of *Titanoides zeuxis*, CNHM P-15551, have associated with them several phalanges. No elements of the forefoot of *Coryphodon* sp. of Paleocene age are known, but, of course, the manus of this genus is represented in the Eocene forms.

Primitive features of the pantodont manus are the large separate centrale present in the pantolambdids and barylambdids, and the unreduced lateral digits; all five digits are of nearly equal size in the latter family. As Patterson has already suggested, the greatest specialization of the carpus in this order occurs in *Titanoides* and *Coryphodon;* and the evidence presented here for claws in *Titanoides,* coupled with the distinctly different elephantine unguals of *Coryphodon,* further demonstrate the high degree of specialization attained in the phalanges of the forefoot in the families to which these two genera belong.

SCAPHOID

The primitive condition of the scaphoid in the Pantodonta is best seen in *Pantolambda bathmodon* as figured by Osborn (1898*b*: fig. 12). In this species the scaphoid is relatively compressed dorsoventrally compared to that of the barylambdids, and the surface for articulation with the radius is broader and flatter than in the latter family. Moreover, the scaphoid in the Pantolambdidae and Barylambdidae agrees in showing no fusion with the centrale. As Patterson (1939*a*) has pointed out, the scaphoid of *Titanoides* is somewhat intermediate between the foregoing forms and *Coryphodon,* in that it is fused with the centrale. However, the fused centrale is much more distinct in *Titanoides.*

The scaphoid of *Titanoides* is relatively thick, while the process on its posteromedial face is comparatively small and deeper than it is broad. This process in the barylambdids, on the other hand, is shallow dorsoventrally and forms a broad semicircular wing extending from the anteromedial corner of the bone posteriorly. In *Coryphodon* the fusion and reduction of the centrale is virtually complete. It is represented by a small process located at the distal margin of the external surface of this bone between the trapezoid and lunar. Further, in this form the tubercle on the medial side of the bone is better defined, being considerably more elongate than in either *Titanoides* or *Barylambda,* and it is rotated somewhat more posteriorly.

LUNAR

The external face of the lunar in the Pantolambdidae, Titanoideidae, and Barylambdidae is rather similar in appearance, being almost as high as it is broad from side to side; but in *Coryphodon,* as Patterson has previously noted, this bone is much flatter, and the dorsal surface for articulation with the radius is proportionately much larger, covering approximately three-fourths of the dorsal surface of the bone. The facet for articulation with the radius in all known pantodonts is noticeably convex. In none of the Paleocene pantodonts do there appear to be distinct facets for the articulation of the lunar with the cuneiform and scaphoid; but in *Coryphodon,* at least in some species, there is an articular face between the scaphoid and lunar. The plantar face of this bone in the Paleocene pantodonts bears two facets, a medial one for articulation with the magnum and a lateral face which looks outward and downward and which contacts the unciform. Both of these surfaces are concave, and the facet for articulation with the magnum in the Paleocene pantodonts is much deeper posteriorly and more strongly concave where it articulates with the dorsointernal surface of the magnum. In the barylambdids, particularly *Leptolambda schmidti,* PU 14879, and 14996, the surface for articulation with the unciform is considerably larger and more expanded in its plantar portion than in *Titanoides,* as can be seen in CNHM P-15520.

CUNEIFORM

The cuneiform of *Pantolambda bathmodon* is incompletely known, but it appears to resemble that of the barylambdids in general proportions. The facet for articulation with the unciform is broadest beneath the medial portion of the bone and in all forms is slightly concave. The posteroexternal portion of this facet is directed downward and tapers rapidly toward the lateral end. In the Torrejonian and Tiffanian pantodonts the facet for contact with the pisiform is situated on the posterodorsal face of the bone. This face is slightly concave, but gently rounded on its posteromedial margin. On this margin the pisiform facet contacts, along its anteroexternal edge, the rounded, concave ulnar facet. In *Titanoides* this bone is considerably broader from front to back along the medial side than in *Leptolambda* and *Barylambda*. *Coryphodon* usually has a shallower medial extension of the cuneiform than in the earlier species. In most other respects the appearance of the cuneiform is remarkably similar in the various pantodont species.

PISIFORM

The head of the pisiform of *Pantolambda bathmodon* is missing in the specimens of this bone I have seen, but judging from the base and anterior portion of this bone, preserved with the left manus of AMNH 2546, there was a distinct constriction posterior to the facet for the ulna. Except for its greater slenderness, the bone is somewhat like that of *Coryphodon* in this respect. The pisiform of *Titanoides* is not known. There are two isolated pisiforms in the Princeton collections from the Silver Coulee beds of the Polecat Bench formation, which appear, on the basis of comparisons with a fragment of the pisiform associated with *Leptolambda schmidti* PU 14996, to belong to this species; the better preserved of these is PU 16485. They differ from the pisiform of *Coryphodon* in having a pronounced ridge running along the lateral side from the anterolateral tip of the forward extremity of this bone, posteriorly to the head. This ridge has the effect of eliminating the constriction or "neck" of the pisiform and shifting the greatest transverse breadth posterior to the region of the neck rather than its occurring in the region of the head.

TRAPEZIUM

The trapezium of the Paleocene pantodonts is considerably larger than the trapezoid. It has the external face extended most distally at its lateroventral margin, where this bone makes a distinct contact with the second metacarpal. On the dorsal surface of the trapezium there is a face for contact with the scaphoid; and, at an angle to this facet on the dorsolateral side of the bone, a facet for contact with the trapezoid. The *Titanoides* trapezium has on its plantar face a conical tubercle directed posteriorly, which articulates with a pronounced concavity on the ventral face of the scaph-

oid. No such tubercle is present in *Leptolambda*, PU 14996, and the contact with the scaphoid covers a very much smaller area.

As Patterson has pointed out, the trapezium of the type specimen of *Haplolambda quinni* is fused with the first metacarpal. This condition does not occur in other pantodonts.

TRAPEZOID

The pantodont trapezoid is intermediate in size between the trapezium and the centrale. Its long axis runs from the external face to the plantar face and it bears on its ventral surface a concavity for articulation with the second metacarpal. On the dorsal face of the trapezoid in the Paleocene pantodonts there is a rounded facet articulating at its lateral extremity in the barylambdids with the centrale, and in *Titanoides* with the fused centrale region only. In the barylambdids, however, there is a distinct face on the dorsomedial surface of this bone for articulation with the scaphoid.

CENTRALE

Among pantodonts this is a separate bone only in the pantolambdids and barylambdids. Patterson (1939a: 371) has suggested that in *Barylambda* this bone may be secondarily enlarged. The centrale of *Leptolambda schmidti*, PU 14996, has its greatest length running from the external face to the plantar surface, and on its plantar side is produced into a bean-shaped tubercle. On its ventromedial face is a concave facet for articulation with the trapezoid, and on its ventrolateral face two rather distinct facets for articulation with the anterior and posterior extremities of the magnum. The dorsolateral surface of the centrale bears a small, slightly convex surface for articulation with the ventromedial extremity of the lunar. This conformation is much like that of this bone in *Barylambda*.

The centrale of *Pantolambda* has not been recovered; but its presence is indicated by a gap between the surrounding bones when articulated. In *Coryphodon* the centrale is almost entirely lost by combination with the scaphoid.

MAGNUM

The magnum of all Paleocene pantodonts, in so far as it is known, has approximately the same general conformation. On its ventral face this bone exhibits a broad convex facet on the lateral side for articulation with the third metacarpal. In the barylambdids and *Coryphodon* there is also a ventromedial facet for articulation with the second metacarpal, with its long axis running from the external to the plantar face of the bone; but in *Titanoides* this surface is relatively smaller and appears to be restricted to the posteromedial extremity of the magnum. In other words, this facet in *Titanoides* is situated on the medial side of the enlarged distopalmar portion of the bone. Above this distopalmar enlargement in *Barylambda, Leptolambda,* and *Titanoides,* as well as in *Pantolambda bathmodon,* there is a knoblike heel with a convex proximal face for articulation with the base of the lunar. This articular heel is situated at the proximopalmar extremity of the bone. The external face of the magnum in *Titanoides* and *Leptolambda* somewhat resembles a triangle with its apices rounded off. In *Titanoides* this face is slightly smaller comparatively than in the other Paleocene members of the order and slightly more oval in outline.

UNCIFORM

The pantodont unciform is by far the largest bone of the carpus and, like the magnum, does not differ strikingly among pantolambdoids. It is deepest at about two-thirds of its transverse diameter from the lateral corner. On its dorsolateral face pantolambdoid unciforms bear a slightly concave facet for articulation with the cuneiform, which looks backward. This facet along its medial margin meets at a sharp angle a dorsomedial facet for articulation with the lunar. This lunar facet in *Titanoides* is almost flat, but in the barylambdids and *Coryphodon* it is somewhat convex. Beneath the lunar facet, on the ventromedial face of the bone, is a somewhat smaller area for articulation with the magnum, and ventral to this is a narrow band running from the external to the plantar face of the unciform for articulation with the third metacarpal. Lateral to this facet, on the ventral face of the bone, are two gently concave facets for the articulation of the fourth and fifth metacarpals respectively. On the plantar extremity of the pantodont unciform is a somewhat enlarged and rugose tubercle, which is directed downwards and backwards. There are few obvious distinctions in this bone among pantodonts; but in *Titanoides* the lateral portion of the external face is comparatively shallow dorsoventrally, while in *Coryphodon* the depth of this portion of the face is almost as great as that of the medial region of the external face. In this respect, this bone in *Pantolambda bathmodon* and the two larger barylambdids appears to be intermediate in structure between the unciforms of *Titanoides* and *Coryphodon.*

METACARPALS

These bones have been recovered with skeletons of only one pantolambdid species, *P. bathmodon.* Among the barylambdids, metacarpals are known in all species, and these bones were also recovered with the left forefoot of the DeBeque specimen of *Titanoides primaevus,* CNHM P-15520.

Perhaps in conjunction with the slenderness of the limb bones as a whole in *P. bathmodon* the metacarpals are relatively long, and seem most like those of *Titanoides* among the Tiffanian pantodonts. In the latter species the anterodistal faces of the metacarpals bear distinct arcuate depressions at the margin of the articular surface of the proximal phalanges not present in

other pantodonts. In the barylambdids the metacarpals are broader at their proximal and distal extremities than in *Titanoides,* and in *Leptolambda* the fifth metacarpal is particularly broad proximally. The foreshortening of pantodont metacarpals reaches its greatest degree in the various species of *Coryphodon,* and in this genus the proximal and distal, lateral expansion of Mc. II through IV is most pronounced, this breadth gives a distinct constricted appearance to the more slender shafts of these bones. There is no fusion of the metacarpals with other elements of the manus in any pantodont except *Haplolambda quinni,* but in this species Mc. I is fused with the trapezium.

PHALANGES

Pantolambdid phalanges of the forefoot are known only in *Pantolambda bathmodon;* but, presumably in all forms here referred to this family, the phalanges were relatively slender and more elongated than in other pantodonts, as they are known to be in the hind foot of most pantolambdid species. All the larger Paleocene members of this order have secondarily foreshortened phalanges as a consequence of their attainment of subgraviportal stature. In *Titanoides,* however, these bones are extremely broad and deep, and only slightly shorter than those of *Pantolambda.* The median phalanges have on their distal faces deep concave articular surfaces for the claws, which have been found in association with five individuals of this genus (see fig. 6).

Among the barylambdids there are apparently some distinctions in the phalanges which are of taxonomic value. The proximal phalanges of *Barylambda* are approximately twice as elongate along the axis of the digit as are these bones in *Leptolambda.* In the former genus the proportions of the proximal and middle phalanges are approximately the same as in *Coryphodon,* but in *Leptolambda* the proximal phalanges are much shorter than in the Eocene pantodonts. The type specimen of *Ignatiolambda* agrees with *Barylambda* in this respect.

The wide range of differences in the ungual phalanges of the pantodont carpus are particularly significant for understanding the ecologic adaptations of this order. Perhaps the most distinctive ungual phalanges seen in any of the Paleocene Pantodonta are the clawlike unguals of *Titanoides primaevus.* Their presence in this pantodont [*Sparactolambda*] was postulated by Patterson in 1939, as a result of the pronounced articular surfaces for claws observed on the median phalanges of the type specimen of *Sparactolambda looki,* here referred to *Titanoides primaevus.* Patterson further pointed out that in conjunction with the clawed manus, *Titanoides* exhibits a number of other anatomical features which suggest a digging or root-pulling way of life.

The exact functioning of the fissured unguals of the manus of *Pantolambda bathmodon* is more difficult to understand; but presumably these phalanges supported small hoofs, which probably were not fissured. Similarly fissured unguals occur in several other kinds of Paleocene mammals, particularly among the Creodonta and Condylarthra in such forms as *Ectoconus, Periptychus,* and *Dissacus.* The phalanges of *Pantolambda bathmodon* figured by Matthew (1937) show that these bones, at least in the most complete skeleton of the species, are distinct in the fore and the hind foot. The phalanges of the hind foot apparently lack all traces of central fissuring.[31] This absence may be characteristic of the genus, for in *P. cavirictus* the one ungual associated with the pes of AMNH 3963, which is apparently the central one, has no really distinct central fissure. Since the central ungual would be expected to show the greatest fissuring, this specimen strongly suggests that, if any of the unguals were fissured in this species, they were not in the pes.

The barylambdids preserve in the unguals of the manus a suggestion of the appearance of these bones in *Pantolambda bathmodon,* but they are relatively foreshortened, broader and deeper than in the Torrejonian species. To what use these large, broad unguals were put is uncertain for they do not resemble the unguals of any of the living graviportal mammals. The central unguals of the modern tapir are sometimes fissured, but this character does not show in the hoof. In the foreshortening of the metacarpals and phalanges of barylambdids there is a suggestion of the beginning at least of a graviportal adaptation. Moreover, the forearm in *Barylambda* is massive with powerful deltoid and entepicondylar ridges on the humerus which suggest that the forefoot could be easily rotated and possibly used for digging, in which the barylambdid unguals if they supported fairly broad hoofs would have functioned in a manner quite distinct from the claws of *Titanoides.*

PELVIS

The pelvis in the Paleocene Pantodonta is remarkably well known. Among the Torrejonian species, the pelvis of *Pantolambda bathmodon* associated with AMNH 16663 is complete except for the rami of the pubes and ischia in the region of their symphysis (see fig. 16). This specimen has already been fully discussed by Matthew (1937: 177), where he notes its general similarity to the pelvis of *Phenacodus primaevus,* and to that seen in the Periptychidae. The provisionally referred pelvis of *Pantolambda cavirictus,* AMNH Number 2455, which Wortmann (1897) mistakenly assigned to a taeniodont, *Psittacotherium multifragum,* is also relatively complete, lacking only the posterior extremities of the ischia and part of the ventral margin of the acetabulum (see fig. 16).

In the Barylambdidae several pelves belonging to *Barylambda faberi* have been recovered in association with the skeletons of this species from the Plateau Val-

ley beds of the DeBeque formation, and a partial pelvis belonging to *Leptolambda schmidti* is associated with an immature skeleton in the Princeton collections from the Silver Coulee beds of Big Horn County, Wyoming. The pelvis of *Titanoides primaevus* is almost completely known from the DeBeque specimens of this species, CNHM 15520 and 15523. A complete pelvis of *Coryphodon* sp., PU 14685, apparently of Paleocene age, has also been recovered from the Polecat Bench formation of Park County, Wyoming.

These pelves demonstrate that by late Paleocene times a considerable amount of specialization of the pelvis in this order had already taken place. The two Torrejonian pelves figured here (16) of *Pantolambda bathmodon* and the referred specimen *Pantolambda cavirictus* are rather distinct; but they agree in possessing a ridge not seen in other pantodonts on the ventral surface of the ilium which runs anteroexternally from the anterior margin of the region of sacro-iliac articulation. The ilium of *Pantolambda cavirictus* has its anterior margin expanded laterally to a greater degree than in the smaller species. The ilia of this specimen AMNH 2455 differ from those of *Titanoides primaevus* and *Coryphodon* in possessing a depression at approximately the center of the crest of the ilium, nor do they resemble in this respect the barylambdids (see Patterson 1935: 153). In one feature, however, the thickening of the anteroexternal angle of the ilium, there is a slight approximation to the enlarged plate-like process present in this position in *Barylambda faberi* and seen to a lesser extent in *Leptolambda schmidti.* As would be expected from the larger size of *Pantolambda cavirictus,* the transverse diameter of the ilia is considerably greater in proportion to their length than in *Pantolambda bathmodon.* A primitive mammalian feature which this specimen shares with all known pantodont pelves is the great elongation of the symphysis.

The distinctions between the pelvis of *Barylambda faberi* and that of *Titanoides primaevus* have already been pointed out by Patterson (1939). They differ principally in that the anteroexternal angle of the ilium of *Barylambda* is directed somewhat more anteriorly and bears the aforementioned plate-like process. In *Titanoides,* on the other hand, there is no enlargement of the anteroexternal angle, and the crest of the ilium is more gently rounded. In this regard, Patterson believed the pelvis of *Titanoides* makes a significant approach to that of *Coryphodon.* Nevertheless, in the relatively smaller area of the gluteal surfaces, the ilia of *Titanoides* are perhaps more like those of *Pantolambda cavirictus* than the corresponding region of the *Coryphodon* pelvis. In *Coryphodon* the ilia are broadly expanded anteriorly, and the anteroexternal angle is strongly directed backward. The Paleocene pelvis of this form (fig. 16) indicates that all the distinctive features of the *Coryphodon* pelvis had already been achieved in late Paleocene times. The ischia of the Paleocene Pantodonta do not differ markedly in shape;

---

[31] See Matthew (1937: pl. 49).

but in *Barylambda, Leptolambda,* and *Coryphodon* they are relatively large in agreement with the larger size of the pelvis in these graviportal or subgraviportal pantodonts. In all the pantolambdoids the spine of the ischium is constricted at the middle but is comparatively deep, and the tuberosity of the ischium is relatively small. The ramus of the ischium is fused with the ramus of the pubis along an anteroposteriorly elongated symphysis. The pubes are slender compared with the ischia, but they are not often preserved. The iliopectinal eminence is low in pantolambdoids but appears to be more pronounced in *Coryphodon,* as in PU 14685. All the known members of this order possess in the pelvis a distinct cotyloid notch on the ventral border of the acetabulum, and an oval obdurator foramen having an anteroposterior diameter distinctly greater than the vertical. For detailed description of the pelvis of *Barylambda faberi* see Patterson (1935: 153).

FEMUR

Of the Torrejonian pantodonts the femur is known only in *Pantolambda bathmodon,* being represented by several specimens in the Cope collections from the Nacimiento formation at the American Museum of Natural History. No complete femora belonging to *Pantolambda cavirictus* are known, but there are fragments presumably belonging to this species in the Cope collection and at Princeton which suggests that the femur of this species does not differ markedly in construction from that of the smaller species. Moreover, the femur of *Caenolambda jepseni,* which may be of Torrejonian age, is quite like that of *Pantolambda bathmodon,* AMNH 16663, though almost twice as large. The femora of the two latter species agree closely in the size and position of the three trochanters except that the lesser trochanter of *Caenolambda jepseni* is comparatively smaller. In this species also, the intercondyloid fossa appears to be somewhat deeper and the bony expansion of the tip of the third trochanter proportionately smaller than in *Pantolambda bathmodon.* These two forms agree in that the position of the pit for the insertion of the *ligamentum teres* is slightly above the posterointernal face of the head, and they both exhibit a relatively deep trochanteric fossa on the posterior face of the great trochanter. A further similarity between these two pantolambdids is that they share a development, intermediate between that of *Barylambda* and *Leptolambda,* of the crests running from the base of the third trochanter along the lateral margin of the femur to the lateral epicondyle and the great trochanter, respectively. The head of the femur of *Pantolambda bathmodon,* AMNH 16663, if it is correctly attached, does not project farther proximally than the great trochanter—in this respect it is unlike that of *Caenolambda jepseni.*

Considering the great separation in time (and most features of the skeleton), the resemblance of the femora

of *Coryphodon,* including those of the two individuals of Paleocene age in the Princeton collections, to this bone in *Pantolambda bathmodon* is striking. Except for the greater slenderness of the *Coryphodon* femur and the slightly more distal position of the medial condyle, these two pantodonts agree well in the size and position of the trochanters. The Paleocene femora of *Coryphodon* belonging to PU Numbers 14682 and 14685 are too poorly preserved for detailed comparisons, but the third trochanters in both specimens are apparently more elongate than is typical of the Eocene species of this genus.

The femur associated with the skeleton of *Titanoides primaevus,* CNHM P-15520, is much distorted, but it is clear that this bone in *Titanoides* differs from the femora of the other pantodonts in the much higher placement of the third trochanter. This femur agrees moreover, with that of *Caenolambda jepseni* in that the external condyles distal to the common trochlear surface are subequal in size, not as in *Coryphodon* and the barylambdids with a larger and slightly more distally placed medial condyle.

Femora belonging to the family Barylambdidae are known only in *Leptolambda schmidti* and *Barylambda faberi,* and in these two forms are quite distinct from one another. These differences are illustrated in Patterson and Simons (1958: fig. 3). In one respect, however—the absence of a pit for the insertion of the *ligamentum teres*—they agree. This absence of a pit here is a distinction which occurs among Pantodonta only in this family.

The femur of *Barylambda faberi* has all three trochanters well developed but is relatively flattened from front to back so that the width across the proximal and distal ends of the bone is comparatively great. The flattened condition of the femur in this species is further exaggerated by the height and lateral extension of the ridges which connect the base of the third trochanter with the lateral epicondyle and the great trochanter. The development of these lateral ridges in *Barylambda faberi* is comparatively greater than in pantolambdid femora. It is this last feature which constitutes the greatest difference between the femora of *Barylambda* and *Leptolambda;* for in the latter form these ridges are almost non-existent and the third trochanter is vestigial, as can be seen in PU 14879 and 14996 as well as CNHM P-15548 and P-26075. Apparently in *Leptolambda* the lesser trochanter is proportionately smaller than in *Barylambda,* but it is not well preserved in the above-mentioned specimens. The reduction of these trochanters and ridges in *Leptolambda* gives the entire femur a comparatively slender appearance, but, in the general thickness of the shaft and the relatively shallow trochanteric fossa, *Leptolambda* and *Barylambda* are similar. The shallowness of the trochanteric fossa may be a distinctive feature of this family. In both these genera the distal condyles agree in that the articular surface of the lateral condyle is broader transversely and shorter anteroposteriorly than the articular surface of the medial condyle. This disproportion of the condyles is not as marked in other pantodont species, but the closest agreement with the barylambdids in this respect is in *Coryphodon*.

PATELLA

The patella is known in *Pantolambda bathmodon, Pantolambda cavirictus, Caenolambda jepseni, Titanoides primaevus,* the two larger barylambdids, and in the Paleocene specimens of *Coryphodon* in the Princeton collections. These patellae are of two basic types. In the Torrejonian species and the barylambdids, they are oval in general outline, with a convex anterior surface. The internal face consists of two connected planes for articulation with the trochlea and is slightly concave and broader at its proximal end. In *Titanoides* and *Coryphodon,* which exhibit the second type, the general form of the body of the patella is similar to that of the first type; but the proximal apex of the bone is situated somewhat more internally and the distal extremity tapers to an elongate and slender point. In *Titanoides,* CNHM P-15451, the patella is poorly preserved and its inner face damaged; but in the specimen of *Coryphodon* sp., PU 14685, of Paleocene age, the medial of the two posterior faces is broader and shorter than the lateral posterior face. The patella of *Caenolambda jepseni,* PU 14863, is somewhat intermediate between the two aforementioned types of pantodont patellae in that it possesses a slender distal extremity which is not, however, as broad as in *Coryphodon* or *Titanoides* and which has a somewhat more oval outline. Moreover, this specimen is distinctive among pantodonts in possessing a pronounced tubercle, situated at the proximal internal apex of the bone. The posterior faces of the patella of *Caenolambda* are also unusual in that they are not as anteroposteriorly elongated as in the later pantodonts, and there is a pronounced fossa beneath and between them on the internal face of the bone.

TIBIA

With the exception of *Ignatiolambda,* the tibia is known in all of the North American Paleocene genera of pantodonts, assuming that the provisionally referred tibia, PU 16481, does actually belong to *Haplolambda.*

This bone in *Pantolambda bathmodon* is represented by several specimens at the American Museum of Natural History, and in this species it is distinct from those of all the later pantodonts in the degree of flattening of the shaft transversely, the height of the cnemial crest, and the retroflexion of the shaft. In the latter feature it makes a distinct approach to the tibia of *Ectoconus* and to this bone in some other Paleocene condylarths. The tibia of *Caenolambda jepseni* is more like that of *Pantolambda bathmodon* than it is like this bone in any of the later pantodonts. The *Caeno-*

lambda tibia resembles that of the smaller pantolambdids, particularly in the transverse flattening of the shaft and in the height of the cnemial crest, but differs from it in lacking retroflexion, and in having a relatively shorter shaft with respect to the transverse diameter of the bone. The medial malleolus of *P. bathmodon* is comparatively much less developed than in *C. jepseni* or, for that matter, than in any of the later pantodonts.

In the barylambdids the shaft of this bone, like that of the pantolambdid tibia, is deeper from front to back than it is in transverse diameter; but the width of the shaft from side to side is comparatively greater than in the pantolambdids. In proportion to the width of the proximal and distal ends of the bone the shaft is still shorter than in *Caenolambda,* and in *Barylambda faberi* the bone is particularly broad from side to side across the proximal condyles and the distal end. The cnemial crest in the latter species is much more distinct than it is in *Leptolambda schmidti* or the referred tibia of *Haplolambda,* but in contrast to the pantolambdids this crest is not made distinct by its height but rather by two shallow grooves running on either side of it, on the anterior face of the bone, which in this species is relatively flat. In this connection it should be pointed out that the isolated tibia and fibula CNHM P-14904, which Patterson initially referred to *Barylambda faberi* but which he later considered too small to belong to this species (see Patterson 1939a: 365), nevertheless agree in general proportion and the above-mentioned features of the tibia with *Barylambda.* This tibia does not resemble, except in its smaller size, any of the tibiae known to belong to *Leptolambda,* or the referred tibia of *Haplolambda.* If the features of the *Barylambda* tibia which are here considered to be distinctive are actually restricted to this genus (and there is no reason to believe they are not), then this tibia and fibula, CNHM P-14904, may eventually prove to belong to a smaller species of *Barylambda* which is otherwise unknown. To refer this specimen to the species *B. faberi* would connote a variability in the size of this bone of almost forty per cent. Some modern mammalian species do occasionally include extremes of variability even greater than this, but there is no other evidence that the size variability of the Paleocene pantodonts was this great.

The tibia of *Leptolambda schmidti* associated with PU 14996, as well as those belonging to this species at the Chicago Museum, differs from that of *Barylambda faberi* in lacking a pronounced cnemial crest and in having a less broad distal extremity in proportion to the transverse diameter of the proximal condyles. In overall appearance the bone is consequently slightly more slender and smaller than in *B. faberi.* These differences between the tibiae of the two larger barylambdids are illustrated in Patterson and Simons (1958: fig. 3). The provisionally referred tibia of *Haplolambda* agrees with *Leptolambda* in all these respects, but its distal extremity is damaged and some comparisons are not possible.

The tibia of *Titanoides primaevus* recovered with CNHM P-15520, is not well preserved, nor does it have any particularly distinctive features. The development of the cnemial crest, however, appears to be somewhat more pronounced than in the other Tiffanian and later pantodonts.

In the Paleocene skeleton PU 14865, referred to *Coryphodon* sp. indet., the tibia resembles most closely that of *Leptolambda schmidti*, except that the distal extremity is still narrower in transverse diameter than in that species, and the development of the cnemial crest is minimal, although Patterson (1939*b*: 108) indicates that there is some variability in the development of this crest, among different species of the genus *Coryphodon.*

The tibiae of all the Paleocene pantodonts agree in having a relatively low spine situated near the center of the posterior margin of the proximal articular surfaces of this bone; in having a low rounded tubercle for the attachment of the patellar ligament at the proximal base of the crest; and in having a rather massive, high medial malleolus elongated from front to back at the distal apex of the tibia.

FIBULA

The fibula is seldom recovered with Paleocene pantodont skeletons, and the articular regions are almost never preserved intact. However, specimens of this bone belonging to *Barylambda faberi* and to *Leptolambda schmidti* are included in the collections of the Chicago Museum. Patterson (1934: 90) gives a full discussion of an isolated fibula CNHM P-14904, which probably belongs to *Barylambda* (see discussion under tibia). A distal portion of the fibula of *Leptolambda schmidti* was recovered with the partial skeleton of this species, PU 14879; and there have also been found much-damaged fibulae belonging to *Pantolambda bathmodon,* AMNH 16663, *Titanoides primaevus* CMNH P-15520 (?), and to the Paleocene specimen referred to *Coryphodon* PU 14682. A fragmentary fibula is also associated with the type of *Caenolambda jepseni* PU 14863.

Neither of the two pantolambdid fibulae mentioned above preserves the proximal or distal articular facets, but the shaft of the fibula of *Pantolambda bathmodon* is much longer in proportion to width than in *Caenolambda jepseni.* In the latter species the expansion at the distal and proximal ends is considerably greater than in the fibula of *Pantolambda bathmodon.* The fibula of *Caenolambda* is almost diamond shaped in cross section and bears sharp ridges running along the external and internal faces of the bone. In *Pantolambda bathmodon,* these ridges are present but are not nearly as distinct, so that in this species the cross section is more rounded. The head of the fibula is not preserved in *P. bathmodon,* but what remains of

the extremities of this bone in *Caenolambda* suggests that as in *Barylambda* the head and distal end were of about the same size.

Patterson (1934: 90) has described the barylambdid fibula from a specimen CNHM, Number P-14904 (see above under tibia), presumably belonging to *Barylambda,* and the later finds of *Barylambda* skeletons included fibulae of this same general sort. In all these specimens there is an articular surface for both the tibia and astragalus, but apparently not for the calcaneum. The fibulae of *Leptolambda,* belonging to PU 14879 and CNHM P-26077, are very similar in size and in comparable proportions to the fibula recovered with the skeleton of *Barylambda faberi* CNHM P-14944. The Paleocene specimen of *Coryphodon* from the Polecat Bench formation PU 14682 includes a much shattered fibula, which does not differ from those of the barylambdids except in the slightly smaller size of the external malleolus and of the proximal articular surface.

PES

The hind foot of the Paleocene Pantodonta is remarkably well known, and for several species more than one tarsus has been recovered. Matthew (1937: pl. 49) illustrates a complete left pes of *Pantolambda bathmodon,* belonging to the nearly complete skeleton of this species, AMNH 16663. He discusses in some detail the primitive nature of this pes and its close resemblance to the hind feet of *Periptychus* and *Ectoconus.* It was principally due to the striking similarities in the pes of *Periptychus* and *Pantolambda bathmodon* that Matthew considered both these forms to belong to the same order, the Taligrada. Gazin (1953: 97) has stressed the similarity of the pes in the tillodonts with Matthew's "taligrade" foot, and it is hard to escape the conclusion that this type of foot which occurs in so many Paleocene subungulates is indicative of a rather close affinity among these orders. Gazin's further observation that the pes of *Trogosus* is definitely unguiculate does not appear to be as significant a distinction from the Pantodonta since the discovery of the clawed manus of *Titanoides.*

Among the pantolambdids the pes of *Pantolambda bathmodon* is completely known, and for *Pantolambda cavirictus* a relatively complete pes, AMNH 3963, exists, which was originally described by Cope. This specimen was more fully discussed and was figured by Osborn in 1898. For *Caenolambda jepseni* much of the pes is preserved with the type specimen, PU 14863. In the Titanoideidae the tarsus is known in *Titanoides primaevus* CNHM P-15520, and for the barylambdids complete hind feet are known in both *Barylambda faberi* and *Leptolambda schmidti.* No complete pes of *Coryphodon* of Paleocene age is known but the elements of the pes associated with the two specimens from the Polecat Bench formation, PU 14682 and 14685, cannot be distinguished from comparable parts of the pes in the Eocene species of this genus.

ASTRAGALUS

The astragalus in all Paleocene pantodonts approximates the coryphodontid type, with no distinct neck and with the distal portion considerably deeper than the proximal. In all these forms there is a lateral facet for the articulation of the fibula and a broad tibial facet which faces vertically. Beneath, the sustentacular and ectal facets are separated by a pronounced groove which ends short of the proximal margin of the plantar face.

The most primitive condition of this bone is presumably that seen in the pantolambdids which still retain a slight suggestion of a distinct astragular head and neck. In *Pantolambda bathmodon* the tibial facet is limited distally by a depressed area containing a large median foramen separating it from the navicular and cuboid facets. A similar depression is present in the astragalus of *Pantolambda cavirictus* AMNH 3963, as well as in this bone in *Caenolambda*. Moreover, in the latter form, the depressed area between the tibial, navicular, and cuboid facets is relatively shallower. The facets for the articulation of the cuboid and navicular in the astragalus of all three pantolambdid species agree in having a relatively shorter vertical depth than occurs in the astragalus of the Tiffanian pantodonts, both barylambdid and titanoideid. As in the astragalus of *Coryphodon* and *Titanoides*, there is a distinct astragalar foramen in pantolambdid astragali. In *Caenolambda*, at least, this is defined posteriorly by a relatively narrow bar of bone and the foramen makes a broad opening into the groove between the plantar facets. *Caenolambda* is not particularly distinct from *Coryphodon* in this respect, although according to Patterson (1934: 92) the development of the astragalar foramen is somewhat variable in the latter form. The facet for the fibula in astragali of *Pantolambda bathmodon* and *Caenolambda jepseni* is relatively deep dorsoventrally. The pantolambdid plantar facets of the astragalus are much as in the later forms, except that the sustentacular facet does not extend as far backward as in the barylambdids, but ends abruptly at the base of a rather pronounced tuberosity or hook, at the posterointernal base of the bone. This facet in the barylambdids, on the other hand, is elongated posteriorly and extends on to the anterointernal base of the bone. The condition of this facet in titanoideids and coryphodontids is intermediate between that of the pantolambdids and barylambdids. In *Titanoides* and *Coryphodon* the sustentacular facet resembles that of the barylambdids in extending on to the base of the hook at the posterointernal corner of the bone; but, like the condition of this facet in the pantolambdids, it is much less elongated from front to back.

The astragalus in the barylambdids is particularly distinctive (among pantodonts) in that, so far as is known, an astragalar foramen is never present. A second definite advance over the pantolambdids, which is shared by *Coryphodon* and *Titanoides* as well, is the distal extension of the tibial facet. This facet in some individual barylambdids makes direct contact with the facets for the navicular and cuboid, so that the reduction of the neck in these forms is virtually complete. In both *Leptolambda* and *Barylambda* astragali there is some variability in this character, and in some specimens there is a slight depression at the distal margin of the tibial facet, which is filled with numerous minute foramina. An astragalus PU 16480, from the same locality in the Silver Coulee beds of the Bighorn Basin, which has yielded a lower dentition of *Haplolambda quinni*, PU 16445, may belong to this species and possibly to the same individual as the lower dentition. The ventral face of this bone is heavily damaged but in the depth of the fossa anterior to the tibial facet it approximates an intermediate condition between the deep depression in this region in the pantolambdids and the much shallower groove occurring in the larger barylambdids.

The absence of the astragalar foramen in the barylambdids is interesting since the presence or absence of this structure has been considered by some authors a feature of definite taxonomic significance. Presumably its disappearance in the barylambdids is a result of the trend toward an almost fully graviportal stature, particularly in the hind limb of this family. Although *Coryphodon* appears to be more digitigrade than *Barylambda* and is, therefore, more advanced toward a true graviportal condition, complete reduction of the astragalar foramen has not taken place in the former genus. However, assuming that the great weight of the barylambdid pelvic region necessitated the enlargement of the tibial facet and the strengthening of the posterior margin of the astragalus through the crowding out of the astragalar foramen, it is understandable that this would have occurred among the barylambdids more readily than in the coryphodontids, for in proportion to the body as a whole the hind quarters of *Coryphodon* are far smaller than they are in *Barylambda* and *Leptolambda*. The presence of an astragalar foramen in most of the members of this order reinforces the idea that the pantodonts belong to the same general division of the Mammalia as the Creodonta and Condylarthra.

The astragalus of *Titanoides* is rather similar to that of *Coryphodon* particularly in having a comparatively small facet for articulation with the cuboid and in the shape and position of the plantar facets. It differs from that of *Coryphodon*, however, in the distinctly greater concavity of the ectal facets and in a more pronounced extension medially of the ventral base of the fibular facet. Both these pantodonts possess an astragalar foramen as well but in *Titanoides* it is placed somewhat more anteriorly. Furthermore, the astragalus of *Titanoides* differs from that of *Coryphodon* and agrees with this bone in the pantolambdids and barylambdids in having a larger angle between the planes of the navicular and tibial facets.

CALCANEUM

Complete calcanea in the Paleocene Pantodonta are known only in *Pantolambda bathmodon* and in the two larger barylambdids but fragmentary specimens of this bone are associated with *Titanoides*, CNHM P-15520, and a Paleocene specimen of *Coryphodon*, PU 14682.

In the barylambdids the *tuber calcis* is broad and relatively short and deep and the posterior extremity is swollen and rugose much as in *Coryphodon*. In *Pantolambda bathmodon* the heel is considerably more elongate as might be expected from the much smaller size of this species. It, too, is expanded at the posterior extremity. The heel in *Titanoides* is not well preserved but it appears to be somewhat more slender than in the barylambdids and *Coryphodon*. In all these forms the two facets are separated by a shallow groove; the ectal facet is the higher and the sustentacular facet extends forward and somewhat externally where it contacts the cuboid facet at a sharp angle. The latter facet, however, is seldom well preserved.

The barylambdid calcaneum has one striking and apparently distinctive feature in that in both *Barylambda* and *Leptolambda* there is a distinct accessory tubercle projecting internally and slightly posteriorly and arising from the posteroventral base of the sustentacular process. There are few other determinable distinctions in pantodont calcanea. The sustentacular facet of the *Titanoides* calcaneum appears to have a larger area than in the pantolambdids and coryphodontids but agrees with them in its being more planar and less convex than this facet is in the barylambdids.

NAVICULAR

A complete navicular is not known in *Pantolambda* but the fragments preserved suggest that the navicular of this form was quite similar to that of *Caenolambda*, PU 14683. Of the Tiffanian pantodonts, naviculars of the two larger barylambdids are known as well as a fragmentary navicular associated with the Plateau Valley skeleton of *Titanoides*, CNHM P-15520. An isolated navicular from the Rock Bench Quarry, PU 16445, is of the general pantodont type but is quite large for a Torrejonian mammal, being approximately a third larger than the navicular of *Caenolambda jepseni*. This navicular is not particularly like that of any of the other pantodonts, and it is possible that it belongs to an otherwise unknown species. A species which is much the largest known Torrejonian mammal.

The navicular of *Caenolambda jepseni* is strongly arcuate as seen from above, and possesses a rounded and well-developed posteroexternal process. The facet for articulation with the head of the astragalus is concave. At its lateral margin this facet contacts at an acute angle the anteroexternally directed facet for articulation with the ectocuneiform. The three external facets for the

articulation of the cuneiform cover approximately the dorsal two-thirds of the convex outer face of this bone, and the remaining portion of the bone curves around the medial side of the astragalar head. This latter portion of the bone is rugose and bears a ventral ridge which is extended almost as far ventrally as is the posteroexternal process.

In *Barylambda* and *Leptolambda* the navicular closely resembles that of *Caenolambda* except in size; but in the former the bone is perhaps a little shallower dorsoventrally and the posteroexternal process a little less distinct. The navicular of the Plateau Valley specimen of *Titanoides* is poorly preserved but an isolated specimen of a navicular thought to be of *Titanoides* from the Crocodile Tooth Quarry in the Silver Coulee beds, of the Bighorn Basin, PU 16485, which is much better preserved, agrees well with the Plateau Valley specimens. These two bones differ from the pantolambdid and barylambdid naviculars discussed above in being much shorter from their medial to lateral margins in proportion to their dorsoventral depth. The surface for articulation with the astragalus is far less concave and the facets for the three cuneiforms cover over four-fifths of the external face of the bone. The posteromedial flange or wing is proportionately much smaller. Reduction of this flange in *Titanoides* is greater than it is in *Coryphodon*. Another distinction of the *Titanoides* navicular is that the posteroexternal process is shifted medially and is not separated from the body of the bone as in the pantolambdids and barylambdids, but bears on its posterior surface a ventral extension of the astragalar facet and on its external surface an extension of the plane of the facet for articulation with the ectocuneiform.

CUNEIFORMS

The mesocuneiform is known in *Pantolambda bathmodon* and *Caenolambda jepseni* among the pantolambdids, and among the barylambdids in *Leptolambda schmidti* and *Barylambda faberi*. This bone is also preserved with the Plateau Valley skeleton of *Titanoides primaevus*. The mesocuneiform varies less in conformation in the Pantodonta than do the other cuneiforms. In all pantolambdoids the ventral face of the bone is roughly oval, concave, and has its greatest length running anteroposteriorly. In the barylambdids, moreover, the mesocuneiform is relatively thick from front to back and has a greater area for articulation with the navicular than in *Caenolambda* and *Titanoides*. The navicular facet of this bone in the barylambdids is very flat but in *Caenolambda* and *Titanoides* it is gently concave. In all three forms this facet meets, along its external margin, the facet for articulation with the ectocuneiform.

The external face of the mesocuneiform is approximately as broad as high, but in *Caenolambda* it is somewhat wider from side to side. At the plantar tip of this bone in these two forms there is a slight tuberosity. In *Titanoides* this tuberosity is somewhat expanded at its tip. In *Coryphodon* the whole mesocuneiform is much flattened proximodistally and does not extend half as far in the plantar direction as does the ectocuneiform.

The ectocuneiform of *Leptolambda* has a cubical external lobe. Its plantar tip is directed somewhat externally and the navicular facet extends well down the length of the bone toward this tip. In *Coryphodon*, however, the navicular facet stops about the middle of the bone which bears on its plantar extremity a large rugose tuberosity, separated by a distinct neck from the dorsal part of the bone. This conformation of the ecto-cuneiform in *Coryphodon* is not really paralleled by that of any of the Torrejonian and Tiffanian pantodonts, but in *Titanoides* and *Pantolambda cavirictus* the plantar extremity of this bone is more expanded than it is among the barylambdids. As might be expected, in the more graviportal *Coryphodon* the ectocuneiform is compressed proximodistally, so that the greatest dimension of the external face is from side to side, just the reverse of the condition in *Pantolambda* where the bone is higher than it is broad. In *Titanoides, Lepto-lambda,* and *Barylambda,* the bone is intermediate in this respect and the two dimensions of the external face are about the same.

The dorsal face of the cuboid is rugose in all the Paleocene pantodonts and is the largest single surface of the bone. At the posteroexternal corner of this face is an oval facet for articulation with the calcaneum, which in the barylambdids is slightly concave, but which in *Caenolambda, Titanoides* and *Pantolambda* is convex. In *Leptolambda* this calcaneal facet is separated from the astragalar facet along its posterointernal border by a low ridge, and the astragalar facet is the larger of the two and is also slightly concave. In *Titanoides,* on the other hand, the astragalar facet is considerably smaller than the calcaneal and is much less concave. The sizes of these two facets in *Coryphodon* and the pantolambdids appear to be more nearly equal to each other. In bary-lambdids the facet for the navicular consists of a narrow dorsoplantar band on the medial surface of the bone situated above a rectangular facet for the ectocuneiform, but in *Caenolambda jepseni* and *Pantolambda cavirictus* there is no facet for the navicular, and these bones appear to be separated by the posterointernal angle of the ectocuneiform. In *Titanoides* and *Coryphodon* there is a narrow contact with the navicular along the external margin of the cuboid which, however, is less distinct than in barylambdids.

The metatarsals of *Pantolambda bathmodon* figured by Matthew (1937: pl. 49) are much slenderer than those of *Pantolambda cavirictus,* AMNH 3963, which are, in turn, slenderer than those preserved with the pes of the type of *Caenolambda jepseni.* The fifth metatarsal of all the Paleocene pantodonts bears a prominent

posteroexternal tubercle which in *Titanoides* and the barylambdids (in conjunction with the marked fore-shortening of the fifth metatarsal) makes this bone approximately as broad as it is long. In all the panto-lambdoids the first metatarsal is the shortest, and in the barylambdids it is, comparatively, the most foreshortened.

The phalanges of the hind foot of the Paleocene pantodonts are not well known, since they have been recovered only in *Pantolambda bathmodon* and *Panto-lambda cavirictus* and in the two larger species of barylambdids. In *Pantolambda cavirictus* the proximal and mesial phalanges are less foreshortened than are those of *Coryphodon,* and in *Leptolambda schmidti* these bones appear to be still more reduced. Otherwise, the proximal and mesial phalanges of the two larger barylambdids approximate closely the condition of these bones in *Coryphodon.*

The ungual phalanges of the Paleocene pantodonts are particularly interesting for their adaptive significance. In contrast to the fissuring seen in the unguals of the fore foot in *Pantolambda,* this condition in the pes is much less pronounced, and in neither the larger nor the smaller species of this genus is there a really distinct central fissure. However, an exceedingly well-preserved ungual associated with the type of *Caenolambda jepseni* is strongly fissured at the tip and does not differ greatly from the fissured unguals of the pes in *Barylambda faberi* and *Leptolambda schmidti,* except that it is comparatively flattened. The unguals of *Titanoides* are not known in the pes but presumably they would have resembled the claws present in the manus of this form, although no doubt, as in the chalicotheres, they would not have been so large. In contrast to the fissured unguals of the barylambdid-pantolambdid type, these phalanges in *Coryphodon* lack any suggestion of fissuring, are flattened and expanded along their anterior border, and resemble those of more distinctly graviportal mammals, such as *Elephas.*

V. GEOLOGIC AND GEOGRAPHIC OCCURRENCE

The Pantodonta have an exclusively holarctic distri-bution, occurring, so far as is known, only in North America, Europe, and Asia. All the earliest species (initially appearing about the middle of the Paleocene) are restricted to North America, but there is no definite evidence that the primary radiation of the order took place on this continent. However, available specimens suggest this. The presence of *Coryphodon* in Europe in the early Eocene, coupled with the absence of species transitional to *Coryphodon* in the North American middle Paleocene radiation, leaves both Eurasia and North America as possible centers for the origin of the coryphodontids. An invasion of *Coryphodon* from the

Old World at or close to the end of Tiffanian time, bringing about the extinction of the Tiffanian pantodonts through competition, might explain the seemingly abrupt disappearance of the Pantolambdidae, Barylambdidae, and Titanoideidae, prior to Clarkforkian time in North America, but *Coryphodon* is nowhere associated with earlier pantodonts. That any of these forms were in direct ecological competition is also most unlikely. It seems improbable that the presence of similar species of *Coryphodon* in North America and Europe in the early Eocene is evidence for a North Atlantic land bridge then, as suggested by Flerow (1957: 80).

In beds of late Paleocene or early Eocene age in Mongolia occurs *Haplolambda planicanina,* the only Old World pantolambdoid. As Flerow suggested, this may be a late survivor of an earlier radiation, which appears to have been North American.

At the end of early Eocene times in Europe and North America the sole remaining pantodont, *Coryphodon,* died out. In Asia on the other hand, two Eocene genera of the order, *Coryphodon* of the middle (?), and *Eudinoceras* of the late Eocene are known; and a third huge pantodont, *Hypercoryphodon thompsoni,* has been recovered from the Oligocene Houldjin gravels of Inner Mongolia in association with the giant rhinoceros *Baluchitherium. Hypercoryphodon* is the last and, at least dentally, the most specialized member of the order.

In North America only one pantodont, *Caenolambda,* from Alberta, has been recovered outside the boundaries of the United States. Furthermore, in this country Paleocene members of the order have been discovered in only five western states: North Dakota, Montana, Wyoming, Colorado, and New Mexico. Of the fifteen species of pantodonts occurring in the Paleocene epoch in North America, five are found in the *upper* Torrejon beds of New Mexico or their equivalents; nine in the Tiffany formation or equivalents; and one in the Clark Fork beds. Evidently, then, the order as we know it had its maximum radiation in Tiffanian time.

The most primitive pantodont species, *Pantolambda bathmodon,* first recovered from the Torrejon beds of the San Juan Basin of New Mexico, was described by Cope (1882). In 1884 he published a description of a larger species *P. caviricus,* also collected in the San Juan Basin. Because both species of this genus are apparently restricted to the upper levels of the Torrejon beds, this part of the Nacimiento formation was named the *Pantolambda* zone by Osborn (1909). Whether the faunal separation between the upper or *Pantolambda* zone and the lower or *Deltatherium* zone of the Torrejon beds is due to a considerable lapse of time between

---

FIG. 7. Hypothetical restorations of some Paleocene pantodonts. (Same scale as the human figure.)

A. *Coryphodon.*
B. *Barylambda.*
C. *Titanoides primaevus.*
D. *Caenolambda.*
E. *Pantolambda caviricus.*
F. *Pantolambda bathmodon.*

TABLE 1

Distribution of Pantodont Species in North America		Torrejonian			Tiffanian							Clarkforkian		Wasatchian		
		Torrejon	Rock Bench	Lebo 2	Tiffany	Silver Coulee	Melville	Plateau Valley	Bison Basin	Fort Union (Bara sec.?)	Scarritt Quarry	Clark Fork	Bighorn Basin (Utah)	Gray Bull	Lysite	Lost Cabin
Pantolambdidae	Pantolambda bathmodon	x														
	Pantolambda intermedius		x													
	Pantolambda caviricus	x	x	x												
	Caenolambda pattersoni							x								
	Caenolambda jepseni	x?														
	Caenolambda sp.											x				
Titanoideidae	Titanoides primaevus					x			x	x	x					
	Titanoides gidleyi					x	x?									
	Titanoides zeuxis								x	x						
	Titanoides mojus					x										
	Titanoides simpsoni				x											
Barylambdidae	Barylambda faberi							x								
	Haplolambda quinni					x	x									
	Leptolambda schmidti					x	x									
	Ignatiolambda barnesi				x											
Coryphodontidae	Coryphodon proterus											x				
	Coryphodon testis etc.													x	x	x	x

? = stratigraphic horizon uncertain.

---

the deposition of the two zones, as Osborn thought, or is the result of a much briefer hiatus or continuous deposition combined with a faunal change, is not clear. Wilson (1956: 83) remarks:

The absence of *Pantolambda* is real enough at localities 9 and 13 [in the *Deltatherium*] zone, but its absence is presumably either controlled by facies or by its being restricted in geographic distribution to more northern areas at the time the strata of the *Deltatherium* zone were being deposited.

P. bathmodon and *P. caviricus* occur together in the Torrejon beds of New Mexico but not in the Torrejonian equivalents of Wyoming or Montana, where the former seems to be absent and *P. caviricus* is rare.[32] However, in the Gidley Quarry in the upper part of the Lebo beds (No. 2) of the Crazy Mountain Field, Montana, the latter species occurs together with a third member of this genus, *P. intermedius.* The rarity of

[32] Douglass (1902: pl. 29, figs. 11, 12) illustrates a single upper premolar PU 13761 from the Crazy Mountain Field, possibly locality 5 or 6, which may belong to *P. bathmodon.* The tooth is about a quarter smaller than the P^2 normally is in this species but is most like it of any of the teeth. The specimen appears to belong to a pantodont but not necessarily to *P. bathmodon.*

Pantolambda in the north is emphasized by the fact that at the Rock Bench Quarry in the Polecat Bench formation, Park County, Wyoming, only one fragmentary tooth and three podials referable to this genus have been found, although this locality has yielded over eleven hundred fragmentary mammalian jaws and teeth. Only two out of twenty-five hundred isolated teeth recently collected by Princeton field parties in the Tongue River Paleocene (Torrejonian) of Carter County, Montana, are pantodont.

Specimens referable to the genus *Pantolambda* have not been found outside the localities discussed above except for a single upper premolar identified as *P. cavirictus* by Gidley, and recovered from the Fort Union formation of Billings County, North Dakota. (See Lloyd and Hares, 1915). The only other pantodont species occurring in Torrejonian levels is *Titanoides simpsoni*, which is represented by a single specimen from the Gidley Quarry. There is, however, a reasonable probability that the type specimen of *Caenolambda jepseni* may also be of Torrejonian age. It was collected from the Polecat Bench formation in Big Horn County, Wyoming, approximately at the boundary between the Rock Bench beds and the Silver Coulee beds which in this area cannot easily be distinguished lithologically. The fauna associated with this specimen includes two forms which have not been found in the Polecat Bench formation above the Rock Bench level, namely *Picrodus* sp. and *Ptilodus wyomingensis*. The *Ptilodus* specimens, however, may be slightly more advanced than those of the Rock Bench Quarry, in that they are perceptibly larger and sometimes have more cusps on the P^4. Nevertheless, these differences are so slight that they are probably not outside the range of the Rock Bench population of *Ptilodus wyomingensis*. The number of specimens of the P^4 of *Ptilodus* from the Rock Bench Quarry is, however, too small to indicate its range of variability. *Picrodus*, on the other hand, also occurs in the lower part of the Melville formation, and its value as an indicator of Torrejonian age is consequently lessened.

The only other certainly identified specimen of this pantolambdid genus, the type of *Caenolambda pattersoni*, was recovered from a level in the Bison Basin which has an associated fauna that correlates with that of the lower Tiffany and Silver Coulee beds. It may be that *C. jepseni* is a slightly earlier species of this genus. Evidently *Caenolambda* occurs approximately at the boundary between these two stages, but correlation of the associated faunas is not accurate enough at present to make its exact range in time certain. Two isolated pantolambdid upper molars from the lower Melville formation USNM Numbers 6155, 9658, probably belong to *Pantolambda cavirictus*. They do not approach in any significant way the upper molars of *Caenolambda*, although the two forms do not differ greatly in the upper dentition. The upper M^2 of *Caenolambda* sp. described as *Pantolambda* sp. by Russell

(1948) from the Saunders series of Central Alberta may also be of this same general age. The latter three molars are the only other pantodont specimens occurring in beds that are in approximately the same stratigraphic position close to the Torrejonian-Tiffanian boundary as those from which the two most complete specimens of *Caenolambda* have been recovered. A possible exception to this observation may be the horizon of the type of *Titanoides majus* which was collected near Sage Point on the Polecat Bench at a very low level in the Silver Coulee beds.

Much greater certainty as to the early Tiffanian age of the large and striking Paleocene pantodonts belonging to the families Barylambdidae and Titanoideidae has resulted from the discovery of many of these large mammals in the Polecat Bench formation of the Bighorn Basin of Wyoming, occurring definitely above the Rock Bench beds with good faunal associations, and clearly distributed throughout the lower several hundred feet of the Silver Coulee beds.[83]

The pantodonts belonging primarily to the genera *Titanoides* and *Leptolambda*, found in the Big Horn County, Wyoming, exposures of Silver Coulee beds, are associated with large faunas which appear to correlate well with those of the early Tiffanian of the Bison Basin of south-central Wyoming (see Gazin 1956).

Inasmuch as the species *Titanoides primaevus, Haplolambda quinni,* and *Leptolambda schmidti* occur both in the Silver Coulee beds of the Polecat Bench formation of Wyoming and in the Plateau Valley beds of the DeBeque formation of Mesa County, Colorado, these forms corroborate Patterson's view that the latter fauna is of Tiffanian age.

The exact age of the beds from which the type specimen of *Titanoides primaevus* was collected in the Sentinel Butte shale member of the type section of the Fort Union formation near Buford, North Dakota, was long uncertain. Brown (1948) had however shown that on the basis of plant and invertebrate remains the Sentinel Butte shale is probably of late Paleocene age. The occurrence of *Titanoides primaevus* in the adequately dated late Paleocene of the Silver Coulee beds of the Polecat Bench formation corroborates this view. The best evidence now available indicates that *T. primaevus* is an exclusively Tiffanian species.

In addition to the North American pantodont localities mentioned above is that of the type of *Ignatiolambda barnesi*. The stratigraphic significance of this specimen for the region of Southern Colorado from which it,

---

[83] The principal quarries and localities that have yielded pantodont materials in the Bighorn Basin are situated to the southeast of the Polecat Bench, as has previously been mentioned, although parts of three individuals of *Leptolambda* were collected on the south side of the Polecat Bench in Park County, Wyoming. Extremely fragmentary remains of several other pantodonts have been recovered from the Polecat Bench proper, and all of these (found in the Silver Coulee beds) were at approximately the level of the Princeton Quarry or below.

as well as the remainder of the Tiffany fauna, was collected makes advisable the inclusion of the following statement written in 1955, by Dr. G. G. Simpson of the American Museum of Natural History. I am indebted to Dr. Simpson for allowing this interesting information to be published here.

The Tiffany fauna of the northern part of the San Juan Basin (La Plata and Archuleta Counties, Colorado) first revealed the presence of late Paleocene beds and fossils in North America. Other and more extensive late Paleocene faunas have since been discovered elsewhere in the Rocky Mountain Region, but the Tiffany remains a standard and is the type of the Tiffanian age and stage (see Wood et al., 1941). Both its fauna and its stratigraphy have, however, remained rather poorly known. The known mammals have included a microfauna of peculiar facies from a single block of clay in the famous Mason Pocket and a few poorly preserved scattered surface specimens from elsewhere. (See Simpson, 1935a, b, c, and references there.) More recently, finds by the Chicago Natural History Museum in the Tiffanian Plateau Valley local fauna of the DeBeque Formation, farther north in Colorado (Mesa County), have suggested that the late Paleocene was especially characterized by a flowering of the peculiar group of archaic ungulates now designated as the Pantodonta. (See Patterson, 1939a, and earlier papers there cited.) This impression has to some extent been corroborated by scattered finds in North Dakota, Montana, and Wyoming. It has, therefore, become somewhat anomalous that the type Tiffanian has included no known pantodonts. The anomaly is removed by the specimen described here (Ignatiolambda).

The specimen was found by Harley Barnes, Jr., while working for the United States Geological Survey. The survey had no immediate facilities for the difficult task of collecting the specimen, and the find was made in a region where the American Museum had already been working. The prospect was therefore generously turned over to us. We are much indebted to Mr. Barnes and to A. D. Zapp not only for this lead, but also for considerable other assistance in the field, and to administrative officers of the Geological Survey for authorizing and facilitating such cooperation. The actual collection of the specimen was done in 1951 by George Whitaker, of our fossil vertebrate laboratory, and by David Kitts and John Ostrom, then student field assistants.

A full discussion of the Tiffany formation with precise designation of all fossil localities will be presented later. It would be premature to go into much detail regarding this one specimen and pending further field work, but the occurrence is unusual and will be summarized here. The locality is Colorado 2 (1951) of the American Museum's San Juan Basin program.[1] It is north of the western part of the Mesa Mountains, between them and the Florida valley, which here swings southwest around the mountains and into the Animas. Here there is a bench held up by a fairly persistent, nearly horizontal, hard arkosic sandstone and fine conglomerate, which locally is an unmistakable horizon marker. The beds above the arkose have fairly typical Tiffany lithology and unquestionably belong to the Tiffany formation.[2] Below the arkose are banded greenish clays with some purple or reddish bands and some sandstone lenses, apparently not andesitic, especially near the top (i.e., near the arkosic horizon marker). These seem to grade downward into the Animas formation without any clear dividing line, although the exposures are not continuous enough to exclude the possibility of an unconformity or other sharply defined change. The beds just below the

arkose could be considered part of the Animas, a lateral facies of the Nacimiento, or a basal portion of the Tiffany.

The arkose, itself contacting sharply below on a clay or siltstone, suggests that this would be a useful formation boundary. But this is true only locally. Elsewhere, notably along the east side of the Animas River and the west and south flanks of the Mesa Mountains, there are other arkoses just as suggestive of a good formation boundary but at various quite different levels. None seems really to be continuous for more than a few miles, and any or all of them may be considered intraformational. That the local "guide" arkose here in question is not likely to represent a valid time boundary or to represent any major change in sedimentation is attested not only by those multiple occurrences of similar arkoses but also by this fossil. It was buried partly in the underlying clay and partly in the arkose, thus extending right across the apparent disconformity, or formation boundary. It is a skeleton, well articulated except parts recently removed by erosion. Therefore the fresh skeleton either was lying partly in and partly on the silt when the arkose was deposited, or it came in with the arkose and sank into completely unconsolidated (apparently practically liquid) silt below. In either case the silt-arkose contact cannot represent any appreciable lapse of time or secular change in local conditions. If this contact were made a formation boundary, we would have a single fossil from two different formations.

As a new genus, Ignatiolambda does not facilitate direct correlation. Similar genera are known only from the Torrejonian and Tiffanian, and the stratigraphic evidence also indicates one or the other of those ages, although it is equivocal as between the two. Ignatiolambda seems to be definitely more advanced than any species currently referred to the Torrejonian genus Pantolambda and to be comparable in this respect with pantodonts of known Tiffanian age. I consider Ignatiolambda as probably Tiffanian in age and a member of the typical Tiffany fauna. Although age is not necessarily pertinent in fixing formational boundaries, this probability suggests the advisability of including the arkose and also the clays and sandstones immediately below it in the basal Tiffany. Their lithology is somewhat different from that of the bulk of the Tiffany, but it is even less like the Animas, Nacimiento, or San José where those formations are definitely recognized. The only other reasonable alternative would be to define a new formation in this part of the section, and that solution is not acceptable from the evidence now available.

---

[1] It is located on aerial photographs, available to those professionally concerned pending publication of a complete locality map, and it has also been noted by Mr. Barnes in his mapping for the U. S. Geological Survey, also unpublished when this was written.

[2] The status of the Tiffany as a rock unit, definable as a formation, has not been generally recognized. U.S.G.S. maps do not distinguish it from the "Wasatch," certainly a misnomer for any formation in this region. I formerly tentatively included it in the San José formation, which is early Eocene, including no Paleocene, in its type area. The previous impression has been that both late Paleocene and early Eocene occur in the Tiffany region, that the two cannot be distinguished lithologically, and that therefore all should be mapped as one rock unit to which the name of the Eocene formation farther south ("Wasatch" or San José) should be extended. It now seems highly probable that there is no Eocene in the Tiffany region and that all these beds are quite distinct from the more southern Eocene formation. If so, they constitute not only a faunal but also a rock unit, which should be called the Tiffany formation.

VI. PALEOECOLOGY

An examination of the sedimentary environments from which Paleocene pantodonts have been recovered might be expected to produce some information as to the type of general adaptation of the various pantodont species. However, no really consistent associations between given species and rock types, such as the occurrence of *Metamynodon* in the channel sandstones of the Oligocene Brulé formation, has been observed. Pantodonts from the Silver Coulee beds of the Polecat Bench formation are recovered in almost all the types of sediments occurring in these beds, including coarse sandstones containing clay pebbles and plant material, relatively pure quartz sandstones, claystones, and mudstones.

Patterson and Simons (1958: 7) have considered the possibility that *Leptolambda* was a more upland animal than *Barylambda*. In the Plateau Valley beds the former occurs only in stream channel sandstones, as if these individuals had been washed in from higher areas. Further, in the Plateau Valley local fauna *Barylambda* is much the commoner form, while in the Bighorn Basin it does not occur at all even though *Leptolambda* (which occurs with *Barylambda* in Colorado) is the most frequently collected pantodont of the Polecat Bench formation. These observations, taken together with the obvious anatomical distinctions between the two genera, strongly suggest that these forms inhabited quite separate environments.

Titanoides specimens collected in the Silver Coulee beds were in most cases recovered from shales, but many isolated teeth and three partial dentitions in jaw fragments were collected from sandstones in two quarries in Big Horn County, Wyoming. In another region, the type specimen of *T. primaevus* was collected in mudstones of the Sentinel Butte shale member of the type section of the Fort Union formation, near Buford, North Dakota. Apparently no ecologic relationship is indicated by the types of sediments in which specimens of *Titanoides* are preserved.

As recently as 1930 only three of the fifteen pantodont species here recognized as occurring in the Paleocene epoch had been described, which suggests the comparative rarity of pantodonts. Such large mammals should theoretically have the best chances of being preserved during fossilization, and their relatively huge bones should facilitate discovery on exposure. Nevertheless, since Cope first described *Pantolambda* in 1882, parts of only about 200 pantodonts of all species have been found in Paleocene beds of Torrejonian, Tiffanian, and Clarkforkian stages, and of these almost half are represented by single teeth only or by small lots of postcranial bones. This total, moreover, is almost certainly disproportionately high because collectors do not often leave even fragmentary material of these unusual mammals in the field. The interest in pantodont remains shown by collectors in the past is evidenced by the number of single pantodont bones in various Paleocene collections.

The apparent rarity of pantodonts may be due largely to the conditions under which burial of mammalian fossils took place during the part of the Paleocene epoch in which they are known to have existed in North America, and not to real rarity at the time. This idea is suggested by the following observation: In one relatively small area of Paleocene sedimentary rock, the Plateau Valley beds of the DeBeque formation in Mesa County, Colorado, pantodonts are common, making up about twenty per cent of the fauna recovered. These beds are also characterized by an absence of microfauna localities from which, because of their large size, most pantodont specimens would naturally be excluded.

The following calculations on the abundance of pantodonts in proportion to total mammalian faunas collected at the same levels suggest what types of sedimentary environments and collection techniques are most likely to yield pantodonts. It should be emphasized that the majority of the figures included here are only approximations.[84]

Tables 2 and 3 indicate that pantodonts are found in high frequencies principally in surface collections and in quarries such as those of the Plateau Valley beds, which include whole skeletons, or those of Big Horn County, Wyoming, where dissociated large bones and teeth of all sorts are found. Apparently the conditions of sedimentation which concentrated Paleocene mammalian remains in large enough quantities to make present-day quarrying practical, at least in many cases, also acted selectively to bring together specimens of roughly the same size range. In current deposition of sediments and detritus, including bones, on sand bars and mud banks along channels, size assortments of this kind are

TABLE 2

NUMBERS OF PANTODONTS AND OTHER MAMMALS FROM
THE PRINCIPAL TORREJONIAN AND TIFFANIAN
LOCALITIES IN WHICH PANTODONTS OCCUR

Provincial age	Formations and local faunas	Panto-donts	All Mammals	% Pantodonts
Torrejonian	1. Nacimiento, N. Mex. (Torrejon l.f.)	54	2,564	2.1
	2. Upper Lebo (No. 2), Montana	15	960	1.6
	3. Polecat Bench, Wyo. (Rock Bench l.f.)	4	1,100	0.36
	4. Carter Co., Mont. (Medicine Rock l.f.)	2 (teeth)	2,550 (teeth)	0.08
		0 (Jaws)	86 (Jaws)	0
Tiffanian	1. Melville, Montana (inc. Scarritt Quarry)	7	2,394	0.29
	2. San José, Colo. (Tiffany l.f.)	1	112	0.89
	3. DeBeque, Colo. (Plateau Valley l.f.)	46	210	21.9
	4. Fallon Co., Mont. (7-Up Butte l.f.)	2	16	12.5

[84] Data on total numbers of mammals recovered from Paleocene localities other than those of the Polecat Bench formation are derived principally from Patterson (1949: 270-271) and Wilson (1956: 82).

TABLE 3

NUMBERS OF PANTODONTS AND OTHER MAMMALS FROM QUARRIES
AND SURFACE LOCALITIES IN THE SILVER COULEE BEDS
(TIFFANIAN) OF THE POLECAT BENCH FORMATION
OF PARK AND BIG HORN COUNTIES, WYOMING

County	Locality	Panto-donts	All Mammals	% Pantodonts
Park County	1. Princeton Quarry	0	594	0
	2. Fossil Hollow loc. (Surface finds)	5	65	7.7
	3. Miscellaneous other localities (largely surface finds)	12	228	5.3
Big Horn County	1. Crocodile Tooth Quarry	30	77	38.9
	2. Cedar Point Quarry	7	170	4.1
	3. Jepsen Quarry	5	27	18.6
	4. Divide Quarry	4	27	14.8
	5. Ant Hill Quarry	2	80	2.5
	6. Miscellaneous other localities (largely surface finds)	10	45	22.2

often produced. Similar conditions in the Paleocene would explain why no pantodonts were recovered in the large fauna of the Princeton Quarry, while surface collecting at the Fossil Hollow locality at approximately the same stratigraphic level has yielded tooth and jaw fragments of several pantodonts including both *Titanoides* and *Leptolambda*.

Simpson (1937a: 30) has pointed out that of the hundreds of mammalian fossils collected in the Crazy Mountain Field of Montana less than a dozen were associations of upper and lower dentitions and that: "Nothing approaching a complete skeleton has ever been found." A similar situation obtains in the collections from the Polecat Bench in the Bighorn Basin, although associated dentitions are not quite so uncommon as they are in the Crazy Mountain field. Only one pantodont from this region of the basin (Polecat Bench proper) has associated upper and lower dentition and postcranial bones. This is a specimen of *Leptolambda schmidti*, PU 14680, in which much of the anterior half of the skeleton is preserved.

To the southeast of the Polecat Bench in the Bighorn Basin between Lovell and Greybull in Big Horn County, Wyoming, the Paleocene deposits have yielded many more isolated occurrences of pantodonts as well as higher percentages of these mammals in the quarries. This high rate of recovery of pantodonts suggests that either the environment or conditions of sedimentation were somewhat distinct in this region of the basin in Tiffanian time.

A striking feature of the unassociated pantodont teeth from Crocodile Tooth Quarry in the SW ¼ of Sec. 5, T 54 N, R 95 W, Big Horn County, Wyoming, is that most of the sixteen separate teeth of *Titanoides* and seven of the *Leptolambda* type exhibit an erosion of the dentine of the roots circling the bases of the teeth in the region of the edge of the enamel of the crown. The possibility that this erosion might be due to diseases of the mouth in these individuals has been suggested. Such erosion of the enamel does occur sporadically in man

and can resemble the condition seen in these teeth. Nevertheless, it is unlikely that such a high percentage of the pantodont teeth recovered from this one locality could have been diseased, and for this reason, and because erosion is also present in dentitions of phenacodonts and arctocyonids from the same quarry, it is more likely that the erosion of the enamel was brought about by *post mortem* action of some sort, perhaps by chemical solution after burial, or just possibly in crocodile stomachs where the teeth may have served as "gastroliths."

A much more definite case of disease in the Pantodonta is indicated by the apparent effects of actinomycosis present in the mandible of the type specimen of *Barylambda faberi*, CNHM P-14637 (see Patterson 1933: 421). If the tumorous mass on the lingual margin of the left horizontal jaw ramus of this specimen has been properly diagnosed, then it would corroborate the supposition that this pantodont was, at least in part, herbivorous. This is due to the fact that *Actinomyces* fungi are acquired almost exclusively from feeding on plant stems and blades where these organisms normally live.

The major evidences of adaptations, and therefore ecology, of all fossil forms are most accurately derived from correlations of ecologic adaptations in living animals and plants with their reflection in the structure of the skeleton or hard parts of present-day forms. Those features which can thus be determined are then used to clarify the significance of similar or identical structures in extinct organisms. The study of pantodont adaptations in this way is paradoxical, in that some structures, such as the clawed feet of *Titanoides*, are surprisingly easy to interpret, while other aspects of the anatomy of the group manifest adaptations so distinct from any living mammal as to make satisfactory explanation difficult. With regard to *Barylambda faberi* in particular, Scott (1937b: 480) well summarizes the difficulties of anatomical interpretation as follows:

Like the Ancyclopoda of the Northern Hemisphere, and the Pyrotheria and Astrapotheria of South America, they [*Barylambda* and *Titanoides*] defy interpretation of their mode of life, because they are so totally unlike anything now living and so completely without relatives in the modern world, that little help can be found in analogies.

In spite of this rather formidable warning, one is tempted to suggest hypotheses as to the adaptive functioning of skeletal structures in the various species of Paleocene pantodonts. Several possible adaptations of the pantodont anatomy have already been suggested where they were relevant to the foregoing taxonomic and anatomic discussions. The principal one of these features is the postulated digging function of the fore limb in *Titanoides*. Although there are no living herbivores that subsist mainly by clawing out vegetable matter, there are many living mammals such as *Rangifer* and other artiodactyls and perissodactyls which do so on occasion. Moreover, this way of life has been suggested for several extinct forms including the clawed notoungulate *Homalodotherium*, the perissodactyls belonging to the Chalicotherioidea, and the primitive artiodactyl *Agriochoerus*. The degree to which many of the bones of the anterior part of the body in *Titanoides* have apparently been altered by selection adapting the animal for the greater strains placed on the fore limbs and neck by its chalicothere-like way of life can be clearly indicated by contrast with *Caenolambda,* in which the unguals are not specialized. Although both forms possess enlarged upper canines and similar cheek teeth, the latter type has no suggestion of the enlargement of the cervical vertebrae, thickening of the clavicles and massive development of the coracoid process of the scapula seen in skeletons of *Titanoides*.

Although the other three families of pantodonts are less striking in their adaptations than the Titanoideidae, they are quite distinct morphologically. Scott and Flerow have both suggested that *Barylambda* resembled the ground sloths. Certainly the huge tail and broad pelvis of this form suggest that it could rear back on the hind quarters, but the fore limbs in *Barylambda* are massive compared to those of most of the giant sloths.

Patterson (personal communication) does not believe that there was any similarity to the sloths in adaptation. The structure of the fore limb and especially of the manus argues strongly against such a possibility in his opinion. He suspects that *Barylambda faberi* was a semi-aquatic form, inhabiting soft, rather swampy ground, on which the large feet would have been useful in locomotion. The very powerful, laterally compressed tail could have functioned as an effective swimming organ. The great motility of the forearm in barylambdids may also be the result of a swimming adaptation.

Pantolambda is relatively less specialized than the foregoing genera and its ecology more vague. Wilson (1956: 83) suggested that *Pantolambda bathmodon* might belong to a riparian fauna. There are some interesting resemblances, which may indicate a similarity of habitat, between the feet of the capybara (*Hydrochoerus*) and those of the pantolambdids (except, of course, for the reduction of the hallux and pollux in the former). These resemblances are particularly evident in the similar shape of the unguals, relative length of the phalanges, and sub-equal size of the digits. It would seem that in *Hydrochoerus* and *Pantolambda* the feet, with their small hooves and relatively separate digits, which would splay out and give valuable support in marshy ground, are an adjustment to riparian living. Furthermore, the large hooves of *Barylambda* and *Ignatiolambda* appear to be a more pronounced development of this quality—a development which is to be expected considering that the barylambdids may have arisen from unknown members of the Pantolambdidae.

Coryphodon has frequently in the past been said to have been river-dwelling, and this may be the case,

although the feet are much more elephantine, are adjusted to a subgraviportal habitus, and have more rigid digits than in other families of pantodonts. The phalanges are unlike those of barylambdids and suggest a different way of life, perhaps resembling more that of the hippopotamus.

Frequently the posterointernal base of the lower canine in *Coryphodon* shows a deep groove, which, since it does not encircle the tooth, is not likely to be due to resorption along the gum line. Such grooves are often covered with striations suggesting that these animals pulled up strap- or cordlike foodstuffs that were covered with sand particles (causing the striations). Such worn surfaces have been observed on entelodont canines and by extrapolation from the modern suids attributed to root-pulling feeding habits. This habit would also appear to have occurred in the coryphodonts. One other hypothesis that may be worth mentioning is that, since in some specimens of *Coryphodon* there are very large canines (presumably owing to sexual dimorphism), these animals may have formed small herds defended by larger males. Herding is typical of ungulates and for *Coryphodon* would appear to have been confirmed also by the recent discovery at one site of about a dozen individuals of *Coryphodon* which appear to have been entombed at one time, according to Dr. G. G. Simpson (personal communication).

VII. RELATIONSHIPS TO OTHER ORDERS

The clarification of some of the details of the adaptive radiation of the Pantodonta during the Paleocene epoch emphasizes that by the mid-Paleocene striking modification in the skeleton had already arisen among these primitive mammals. This diversity suggests a long unknown history for the order which Simpson (1945: 241) has categorized as one of "the most distinctive of mammalian groups. . . ." Many general similarities to other early Tertiary orders have been observed, but the great majority of these must be regarded as parellelisms derived from a common eutherian ancestry. Moreover, the range of adaptative distinctions within the order renders uncertain the exact nature of the most primitive stock of this group.

As the result of initial observations by Wood (1923) and later studies by Simpson and Patterson, it has been clearly demonstrated that the Pantodonta and Dinocerata, formerly united in the order Amblypoda, represent distinct groups, and that the correspondence in foot structure between *Coryphodon* and the uintatheres is entirely convergent. In the same way, the great elongation of the upper canines in *Titanoides* can be regarded as an independently developed similarity to the sabre-like canines of the Dinocerata.

General resemblances to the upper molars and lower premolars of pantodonts have been noted in several early Tertiary orders. The simplicity of the basically three cusped upper molar teeth of the Pantodonta permits similarities in conformation to those of other orders to be seen and they are difficult to interpret. Flerow (1952), for instance, considered the dentition of *Haplolambda planicanina* to be basically like that of the Insectivora and Chiroptera and maintained that this pantodont was not herbivorous but insectivorous. It is true that the basically dilambdodont upper molars of the Pantodonta do resemble vaguely those of some insectivores, but similarities of the same degree are certainly not restricted to the unguiculate cohort. The large size of even the smallest pantodonts when compared to insectivores appears to rule out insectivorous feeding habits in the Pantodonta. That the pantodonts and other early Tertiary orders were to a degree omnivorous is, of course, likely.

Matthew (1937: 169) noted a resemblance in the upper fourth premolar of various Artiodactyla, particularly the Anthracotheriidae, to that of *Pantolambda* and *Coryphodon*. As there are few other similarities in the pantodont and artiodactyl skeletons, this agreement is almost surely due to convergence. More significant is his observation that the teeth of *Pantolambda* are like those of the Oxyclaenidae. This remark applies only to *Deltatherium* which Matthew and others, regarded as an oxyclaenid. Comparisons of the basicranium of *Deltatherium* (see Matthew 1937: fig. 14) with that of *Titanoides primaevus*, illustrated here in figure 14, is suggestive of some kind of relationship. Particularly significant is the agreement in both forms in the general position and conformation of the ventral surface of the petrosal. There is in each case a broad, expanded region on the anterointernal surface of this bone which bears a central eminence. This prominence, however, is too large to be the promentorium cochleae, and in *Titanoides,* at least, does not contain the cochlear canals. There is a general similarity in the upper molars of *Deltatherium* and pantodonts, but the premolars of the former are less molarized than those of the Pantodonta. Nevertheless, there is evident within the known history of the Pantodonta a tendency toward the enlargement and complication of the first and second upper premolars. These teeth shift from relative simplicity in *Pantolambda bathmodon,* where they lack the well-developed protocone of the P^3 and P^4, to a stage, exemplified by the upper premolars of *Titanoides* or *Coryphodon,* in which the P^2 has a fully developed protocone, and the P^1 has an incipient protocone similar to that of the P^2 in *Pantolambda.* Projecting this sequence backwards, the hypothetical stage prior to that represented by *Pantolambda* would be one in which only the P^4 had achieved the typical pantodont conformation. Some arctocyonids and even tillodonts exhibit structural approximations to this requirement, though none of the known species of either group are old enough to be at all close to the ancestry of the Pantodonta.

To the many resemblances noted by Matthew (1937) among *Periptychus, Ectoconus, Phenacodus,* and *Pantolambda bathmodon,* should be added the character of retroflexion of the tibia, which occurs among the Pantodonta only in the latter species. Some retroflexion of the tibia is apparent in *Periptychus* and it is present to an even greater extent in *Ectoconus.* As outlined in Section II, Simpson (1937a: 266) has pointed out that the association of the periptychids with *Pantolambda* in the order Taligrada is no longer defensible and has referred the Periptychidae to the Condylarthra. Nevertheless, Matthew (1937) observed some striking similarities between the two groups which should not be overlooked. As Simpson (1937) emphasizes, these forms, particularly the periptychids, cannot be ancestral to the Pantodonta, since they are contemporaries of the earliest genera of the latter order. This situation, however, is unavoidable in all considerations of the phyletic affinities of the Pantodonta, for no earlier forms are known which bear striking or significant resemblances to them with the possible exception of *Deltatherium.* All that can be done is to indicate their similarities to contemporary or later groups and to stress those features which suggest a common ancestry. Most of the more striking similarities of the pantodont dentition to the teeth of other mammals are almost certainly the result of convergence. Among such cases of convergence are the similarities between the lower molars of *Coryphodon* and more particularly *Eudinoceras* and those of some of the macropodid marsupials, particularly *Dorcopsis* and *Macropus.* Moreover, the bilophodonty of the upper molars of *Dorcopsis* also resembles, in a less striking way, the double-crested upper molars of *Eudinoceras mongoliensis* and *Hypercoryphodon thompsoni.* The close correspondence between the lower molars of the tapir and those of *Coryphodon* has been recognized since the time of Owen, but the agreement between *Eudinoceras* lower molars and those of this perissodactyl is even more striking.

De Paula Couto (1952: 371) discussed the similarity in the dentition of the South American Xenungulate *Carodnia,* from beds of Paleocene stages in Patagonia and Brazil, with the teeth of both the Pantodonta and Dinocerata. The lower molars of *Carodnia* are somewhat similar to those of *Coryphodon* and they are still more like those of *Eudinoceras mongoliensis.* The remainder of the dentition of *Carodnia* is less pantodontlike. The manus of *Carodnia* exhibits features of both the coryphodont and barylambdid forefoot. Presumably this resemblance in the two orders with regard to foot structure is largely the result of their both having attained a sub-graviportal stature, and does not indicate close affinity.

Comparisons of the pantodont dentition and skeletal morphology with many other primitive mammals have been made during the course of this study but apart from the observations given above few new similarities of taxonomic significance between the Pantodonta and other mammals have been observed and these are of such a general nature as to be inconclusive. From this negative point of view new emphasis is given to the distinctness of this order. The observation that the talonids of the milk premolars of *Coryphodon* and *Barylambda* are broader than in the permanent dentition may indicate that the relatively narrow talonids of adult pantodont lower molars are a comparatively recent acquisition in the Paleocene members of the Pantodonta.

A. LOGARITHM OF RATIO DIAGRAMS

The logarithm of ratio diagrams included here were constructed in order to make possible easy comparison of ratios of proportion in the lower teeth of various pantodont species. Lower dentitions were used in preference to upper teeth because they are represented in higher numbers of individuals and in more species than are the upper teeth. Teeth in the lower dentition are also more compact structurally, since they lack tapering projections such as the parastyle and metastyle of the pantodont upper premolars and molars, and consequently they are less easily broken. If either parastyle or metastyle on pantodont upper teeth is lost, neither transverse nor anteroposterior diameters can be accurately measured.

Comparisons of measurements of bones of the remainder of the pantodont skeleton (other than teeth) have not been made because distortion in all but a few pantodont crania, mandibles, and long bones makes their measurement unreliable compared to those taken on teeth, which are less often distorted.

For those unfamiliar with the method of construction and use of such logarithm of ratio diagrams the following discussion is included. Additional comments regarding similar diagrams are given by Simpson (1941 : 23).

Through logarithm of ratio plottings of this type it is possible to recognize similarities of proportion with the factor of different absolute size eliminated. If absolute proportions were plotted, then specimens which are identical in form and which differ only in overall size would not, in general, yield paralleling lines, and thus similarities in proportions of measurements would be less apparent. Furthermore, the use of the logarithm differences rather than the ratios themselves allows the plotting of more than two series of measurements of structures in such a manner that any two or all may be quickly compared visually.

A simple example will illustrate this point. If the canines of three species were in the ratio 1 : 2 : 6, and the ratios of the second and third to the first were plotted linearly, with the first represented as the origin, then the distance from the first to the second would be two units, from the first to the third six units; hence the distance from the second to the third would be four units. However, the ratio of the third measurement to the second is not four, but three. On the other hand, if the logs of two and three were marked off, then the distance between the point representing log two to that representing log six would be log three, as desired.

B. MEASUREMENTS OF PANTODONT DENTITIONS

The measurements included in the following pages have been taken as follows: All anteroposterior diameters give maximum length of the tooth near the base,

on its long axis and parallel to the tooth row and jaw ramus, including the anterior and posterior basal cingula if they are present. Moreover, all teeth in each jaw are measured as if they were in a straight row, so that the anteroposterior diameter of the second incisor,

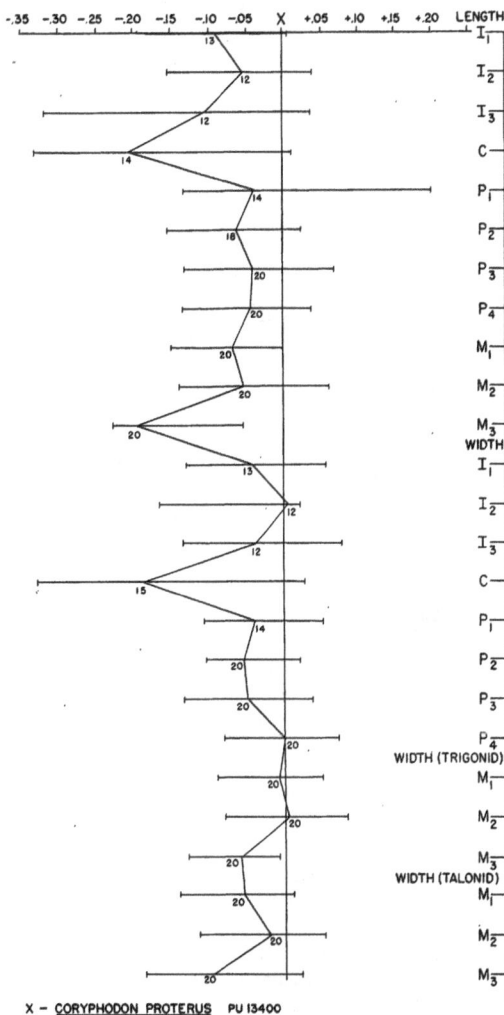

X – <u>CORYPHODON PROTERUS</u>   PU 13400

Fig. 8. Logarithm of ratio diagram of dimensions of lower teeth of *Coryphodon proterus* (standard) as labeled, compared with the same measurements on twenty specimens of *Coryphodon* sp. of early Eocene age. Horizontal lines connect the extreme log differences measured in samples whose size is indicated by the number below the line. The broken vertical line connects the means of these log differences for the Eocene specimens of *Coryphodon*.

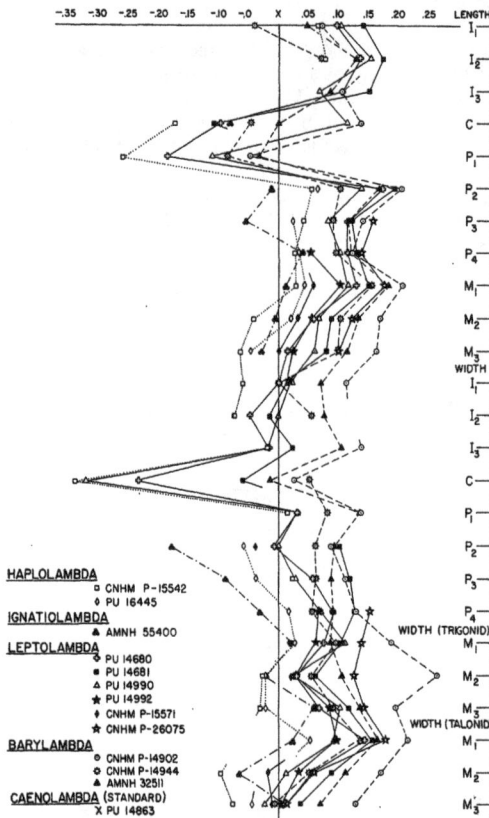

FIG. 9. Logarithm of ratio diagram, plotting lengths and widths separately, of teeth in all known barylambdid lower dentitions and contrasted with the pantolambdid *Caenolambda jepseni* (standard).

An *a* following the measurement indicates a possible error of ten per cent or less. An *e* after a measurement designates that there is sufficient evidence to justify an estimate for the measurement but that its correctness cannot be positively demonstrated.

MEASUREMENTS OF DENTITIONS
All measurements in millimeters

1) *Pantolambda bathmodon*

Upper dentition: (Anteroposterior diameter)	USNM No. 15408	AMNH No. 16663	AMNH No. 16664	AMNH No. 16665
I^1	—	3.7	—	—
I^2	—	3.8	—	—
I^3	—	5.3	—	—
C	—	8.6	9.0	—
P^1	6.0	—	6.1	—
P^2	8.2	8.0	8.2	6.7
P^3	8.8	9.1	8.8	—
P^4	9.0	9.3	9.0	—
M^1	10.7	11.9	11.0a	10.3
M^2	10.9	12.0	11.2	12.2
M^3	8.7	10.0a	10.0a	9.6

(Transverse diameter)				
C	—	—	8.1	—
P^1	4.0	—	4.2	—
P^2	7.2	—	—	7.8
P^3	12.8	—	—	11.8
P^4	13.9	—	—	14.0
M^1	15.1	—	—	16.0a
M^2	17.2	—	17.0	17.8
M^3	17.3a	—	16.5	17.8

Length from anterior margin of P^1 to posterior margin of P^4	29.8	29.5	32.0a	30.0a
Length from anterior margin of M^1 to posterior margin of M^3	30.0	32.2	31.3	31.5
Length from anterior margin of C to posterior margin of M^3..	—	68.0	74.5	73.0a

Lower dentition: (Anteroposterior diameter)	AMNH No. 2552	AMNH No. 3958	AMNH No. 16664	KU No. 8072
C	—	—	—	8.9
P_1	—	—	7.9	6.3
P_2	—	—	—	7.6
P_3	—	—	—	—
P_4	—	8.2a	—	8.2
M_1	—	10.5	10.4	9.5
M_2	10.3	10.6	11.0	10.9
M_3	13.3	14.3	14.5	13.9

(Transverse diameter)				
C	—	— L	—	9.4
P_1	—	—	3.7	3.8
P_2	—	—	—	5.2
P_3	—	—	—	—
P_4	— L	8.1	— L	7.5
M_1, trigonid	—	8.0	—	7.6
M_2, trigonid	7.1	8.2	8.1	8.7
M_3, trigonid	8.0	8.4	8.5	8.4
M_1, talonid	—	8.1	7.8	7.9
M_2, talonid	7.8	8.5	8.4	8.3
M_3, talonid	8.0	7.9	7.2	7.4

for instance, is measured from the part of the tooth nearest the third incisor to the part nearest the first incisor.

Probably the least reliable anteroposterior measurements of pantodont teeth are those taken on the lower premolars and molars, for these teeth usually wear against one another on their anterior and posterior faces, and in some cases this friction has appreciably shortened the tooth lengths. Apparently this somewhat variable wear does not greatly reduce the usefulness of such measurements, for in *Coryphodon* at least, as illustrated in figure 8, the extreme ranges of the ratios of anteroposterior measurements are only slightly greater than those for the transverse measurements.

Widths (transverse diameters) of the pantodont teeth measured are taken along the greatest axis of the tooth at right angles to the tooth row.

2) *Pantolambda cavirictus*

Upper dentition: (Anteroposterior diameter)	USNM No. 21327	AMNH No. 963
I¹	5.6	—
I²	—	—
I³	8.2	—
C	15.2	18.8
P¹	9.1	—
P²	12.0	—
P³	13.2	—
P⁴	14.2	12.9
M¹	17.0	15.7
M²	17.5	15.8
M³	13.0	13.5

(Transverse diameter)

I¹	5.1	—
I²	—	—
I³	7.5	—
C	14.1	14.3
P¹	5.8	—
P²	12.2	—
P³	17.0a	—
P⁴	20.1	17.9
M¹	24.0a	19.8
M²	25.4	22.1
M³	25.7	23.5

Length from anterior margin of P¹ to posterior margin of P⁴..	50.5	—
Length from anterior margin of M¹ to posterior margin of M³	46.0	43.9
Length from anterior margin C to posterior margin of M³...	110.1	106.2

Lower dentition: (Anteroposterior diameter)	AMNH No. 3961 Type
C	16.0a
P₂	11.0
P₃	11.8
P₄	13.0
M₁	17.5
M₂	19.5
M₃	22.4

(Transverse diameter)

C	16.0a
P₂	9.1
P₃	10.8
P₄	11.9
M₁, trigonid	14.5
M₂, trigonid	15.0
M₃, trigonid	13.0
M₁, talonid	14.4
M₂, talonid	14.3
M₃, talonid	12.0a

Length from anterior margin of P₁ to posterior margin of P₄..	60.0e
M₁–M₃	57.0
C–M₃	140.0

3) *Caenolambda jepseni*

Upper dentition: (Anteroposterior diameter)	PU No. 14863 Type
I¹ (?)	6.0
C	28.5

P¹	14.7
P²	14.4
P³	15.0
P⁴	15.0
M²	18.7

(Transverse diameter)

I¹	5.1
C	15.4
P¹	7.3
P²	21.0
P³	24.1
P⁴	24.4
M²	26.2

Lower dentition: (Anteroposterior diameter)

I₁	6.4
I₂	7.4
I₃	9.0
C	16.0
P₁	20.6
P₂	13.8
P₃	15.6
P₄	16.0
M₁	16.6
M₂	20.0
M₃	25.1

(Transverse diameter)

I₁	6.0
I₂	7.0
I₃	7.3
C	15.0
P₁	8.0
P₂	12.5
P₃	14.0
P₄	14.4
M₁, trigonid	14.2
M₂, trigonid	15.8
M₃, trigonid	13.9
M₁, talonid	12.9
M₂, talonid	14.9
M₃, talonid	13.2

Length from anterior margin of P₁ to posterior margin of P₄..	73.0
Length from anterior margin of M₁ to posterior margin of M₃	63.1
Length from anterior margin of canine to posterior margin of M₃	152.6

4) *Titanoides primaevus*

Upper dentition: (Anteroposterior diameter)	PU No. 16490 Type	USNM No. 20029	CNHM No. P-15520 (Type of "S. looki")
I¹	7.0e	—	—
I²	9.2	—	—
I³	10.0	—	—
C	30.0	—	32.4
P¹	12.2	—	12.3
P²	11.0	—	—
P³	11.6	—	11.0
P⁴	12.6	12.7	11.8
M¹	15.6	15.5a	14.5
M²	18.7	18.0	17.8
M³	19.0	19.0	19.5

(Transverse diameter)

I^1	4.0e	—	—
I^2	5.3	—	—
I^3	5.8	—	—
C	19.3	—	15.0a
P^1	9.0	—	8.9
P^2	18.9	—	—
P^3	23.0	—	23.4
P^4	24.0	27.6	—
M^1	22.2	25.0	22.3
M^2	28.2	31.1	28.5
M^3	32.1	36.3	32.4

Length from anterior margin of P^1 to posterior margin of P^4..	57.6	—	60.0
Length from anterior margin of M^1 to posterior margin of M^3	70.9	72.3	71.0a
Length from anterior margin of C to posterior margin of M^3..	163.0	—	—

Lower dentition: (Anteroposterior diameter)	USNM No. 7934 Type	CNHM No. P-15520
I^1	—	10.0
I^2	—	13.8
I^3	—	13.7
P		44.0
P_1	—	13.0
P_2	—	14.4
P_3	16.0a	16.0
P_4	—	17.3
M_1	—	19.6
M_2	27.9	24.0
M_3	32.5	31.0

(Transverse diameter)

I_1	—	6.4
I_2	—	7.8
I_3	—	8.1
C	—	22.1
P_1	—	8.0
P_2	—	11.3
P_3	14.5	14.0
P_4	—	16.2
M_1, trigonid	15.6	15.3
M_2, trigonid	18.8	17.6
M_3, trigonid	19.6	19.7
M_1, talonid	—	13.3
M_2, talonid	17.3	15.3
M_3, talonid	17.0	16.2

Length from anterior margin of P^1 to posterior margin of P^4..	—	57.8
M_1–M_3	—	72.6
C–M_3	—	175.4

5) *Titanoides gidleyi*

Upper dentition: (Anteroposterior diameter)	PU No. 14974
P^2	10.0
P^3	11.0
P^4	11.9
M^1	14.5
M^2	16.3
M^3	17.5

(Transverse diameter)

P^2	18.0
P^3	21.0
P^4	23.4
M^1	22.0
M^2	27.5a
M^3	32.5

Lower dentition: (Anteroposterior diameter)	PU No. 13235 Type
I_2(?)	11.0
C	26.0a
P_2	13.5
P_3	15.2
P_4	18.0
M_1	21.7
M_2	23.2

(Transverse diameter)

I_2(?)	6.5a
C	19.0
P_2	11.4
P_3	11.8
P_4	15.3
M_1, trigonid	15.0
M_2, trigonid	15.7
M_3, trigonid	17.2
M_1, talonid	12.3
M_2, talonid	14.9

6) *Titanoides zeuxis*

Lower dentition: (Anteroposterior diameter)	AMNH No. 35201	CNHM No. P-15551	PU No. 14617
C	—	21.3	
P_1	—	—	12.3
P_2	—	13.0	14.9
P_3	14.8	—	
P_4	—	15.0e	
M_1	—	17.5	
M_2	—	23.8	
M_3	25.1	26.2	

(Transverse diameter)

C	—	14.0	
P_1	—	—	6.9
P_2	—	10.1	10.2
P_3	11.8	—	
P_4	—	12.0	
M_1, trigonid	13.5	13.5	
M_2, trigonid	14.0	14.1	
M_3, trigonid	14.2	14.3	
M_1, talonid	—	11.0	
M_2, talonid	—	—	
M_3, talonid	13.2	12.0e	

7) *Titanoides majus*

Lower dentition: (Anteroposterior diameter)	PU No. 16447 Type
P_2	19.3
P_3	19.5

(Transverse diameter)

P_2	14.2
P_3	15.6

8) *Titanoides simpsoni*

Upper dentition: (Anteroposterior diameter)	AMNH No. 35720 Type
P⁴	11.5e
M¹	13.0a
M²	14.1
M³	11.1

(Transverse diameter)	
P⁴	15.3
M¹	18.5a
M²	20.5
M³	19.6

9) *Haplolambda quinni*

Lower dentition: (Anteroposterior diameter)	PU No. 16445
P₂	16.0
P₃	16.5
P₄	17.2
M₁	18.2
M₂	19.0
M₃	22.5

(Transverse diameter)	
P₂	11.0a
P₃	13.0a
P₄	15.0a
M₁, trigonid	15.2
M₂, trigonid	15.0a
M₃, trigonid	13.3
M₁, talonid	14.5
M₂, talonid	—
M₃, talonid	12.0a

10) *Ignatiolambda barnesi*

Upper dentition: (Anteroposterior diameter)	AMNH No. 55400 Type
C	13.9
P²	10.4
P³	12.3
P⁴	13.8
M¹	17.8
M²	18.0a
M³	17.0a

(Transverse diameter)	
C	12.4
P²	16.7
P³	21.5
P⁴	23.7
M¹	23.9
M²	27.0a
M³	28.0a

Lower dentition: (Anteroposterior diameter)	
C	13.7
P₂	13.6
P₃	13.7
P₄	17.8
M₁	16.7
M₂	20.0
M₃	23.5e

(Transverse diameter)	
C	13.0
P₂	9.2
P₃	11.5
P₄	13.5e
M₁, trigonid	15.0a
M₂, trigonid	15.0a
M₃, trigonid	16.0a
M₁, talonid	13.5
M₂, talonid	13.0a
M₃, talonid	13.3

11) *Leptolambda schmidti*

Upper dentition: (Anteroposterior diameter)	CM No. 11353	CNHM No. P-26075	PU No. 14680	PU No. 14879	PU No. 14990	PU No. 14996	PU No. 16662
I¹	6.0	—	—	—	—	—	—
I²	—	—	7.7	—	—	—	—
I³	8.8	—	9.1	—	—	—	—
C	13.4	14.5	10.9	—	—	—	—
P¹	15.0a	—	14.6	13.8	15.7	—	—
P²	16.0a	18.0a	—	—	—	18.2	18.1
P³	17.0a	20.5	—	17.0	—	17.8	17.3
P⁴	18.0a	—	—	—	—	—	17.0
M¹	26.0a	—	23.0a	—	21.5	24.5	22.0a
M²	27.0a	26.0	23.5a	22.8	24.4	25.3	24.0a
M³	21.0a	21.0	17.4	18.1	—	19.8	18.0e

(Transverse diameter)							
I¹	5.2	—	—	—	—	—	—
I²	—	—	6.3	—	—	—	—
I³	9.0	—	6.8	—	—	—	—
C	13.5	15.0	8.8	—	—	—	—
P¹	14.0	—	8.4	8.0	9.2	—	—
P²	26.0	25.0	—	—	—	25.5	23.8
P³	29.5	27.2	—	25.0	—	27.0	26.6
P⁴	32.0	—	—	—	—	—	28.0
M¹	36.0a	—	31.5	—	33.2a	31.6	28.0
M²	39.0a	36.0	32.5	34.1	35.0a	35.3	32.0e
M³	25.0a	34.5	30.0a	31.2	—	33.4	29.0e

Lower dentition: (Anteroposterior diameter)	PU No. 14680	PU No. 14681	PU No. 14990	PU No. 14992	CNHM No. P-15571	CNHM No. P-26075
I_1	8.0	8.8	8.0	—	—	—
I_2	10.0	11.0	10.5	—	—	—
I_3	11.5	12.7	10.5	—	—	—
C	12.8	15.7	12.3	—	13	—
P_1	13.5	—	16.0	—	—	—
P_2	20.5	21.5	19.0	—	19.0a	—
P_3	20.5	20.5	19.0	—	—	22.5
P_4	21.0	21.6	20.2	18.0	—	22.0
M_1	22.5	23.2a	21.8	21.0	19.0	25.0
M_2	23.0	24.6	23.3	22.8	21.5	26.7
M_3	26.0	30.4	29.0	26.1	25.0a	31.5

(Transverse diameter)						
I_1	6.0	6.2	6.2	—	—	—
I_2	6.2	6.8	7.0	—	—	—
I_3	7.0	7.7	7.0	—	—	—
C	8.8	11.4	7.2	—	8.0	—
P_1	8.6	—	8.6	—	—	—
P_2	12.3	15.7	12.6	—	11.4	—
P_3	16.0	18.3	15.0	—	—	18.0a
P_4	18.0	19.5	17.0	17.0	—	20.5
M_1, trigonid	17.0	18.0	18.2	16.5	15.0	19.5
M_2, trigonid	17.0	18.2	17.2	17.0	17.0	21.0a
M_3, trigonid	16.3	18.2	17.5	17.0	16.0	19.0a
M_1, talonid	17.8	18.3	16.0	16.2	—	19.5
M_2, talonid	17.0	18.3	15.4	16.2	14.4	17.0
M_3, talonid	13.4	14.4	12.5	13.4	13.0	13.5

12) *Barylambda faberi*

Upper dentition: (Anteroposterior diameter)	CM No. 8990	AMNH No. 32511	CNHM No. P-14944	CNHM No. P-14955	CNHM No. P-15075	CNHM No. P-25617
I^1	—	6.5	—	—	—	—
I^2	—	9.2	7.9	—	—	—
I^3	—	11.3	12.0	—	—	10.3
C	21.0a	—	18.8	—	—	18.0a
P^1	—	18.0	15.8	—	—	18.0
P^2	21.7	18.0	17.8	19.5	18.0	—
P^3	22.2	18.9	18.0	20.0	—	19.0e
P^4	22.0	19.0	18.0	24.0e	20.0	20.0
M^1	28.8	27.2	25.0e	31.0	26.5	—
M^2	28.6	27.7	26.0a	30.0	27.4	29.0
M^3	24.3	21.1	20.0	24.0a	22.5	22.0a

(Transverse diameter)						
I^1	—	5.0	—	—	—	—
I^2	—	6.5	9.6	—	—	—
I^3	—	12.1	12.7	—	—	11.0
C	20.0a	—	18.5	—	—	17.0a
P^1	—	14.1	11.0	—	—	12.1
P^2	30.5	28.4	23.2	30.2	31.5	—
P^3	31.5	30.8	29.0	33.0	—	32.1
P^4	37.7	34.0	30.5	36.0	36.8	31.8
M^1	35.0	35.6	31.0e	37.0	37.8	36.0
M^2	41.9	38.8	33.0e	41.5	42.5	40.5
M^3	41.8	35.3	33.5	38.2	38.8	36.0

Length from anterior margin of P^1 to posterior margin of P^4..	75.0e	70.0e	75.0	—	—	—
Length from anterior margin of M^1 to posterior margin of M^3	83.5	73.7	75.0a	—	72.5	—
Length from anterior margin of C_3 to posterior margin of M^3	193.0	161.0	165.5	—	—	—

Lower dentition: (Anteroposterior diameter)	AMNH No. 32511	CNHM No. P-14637 Type	CNHM No. P-14902	CNHM No. P-14944
I_1	7.1	—	7.5	5.8
I_2	10.1	—	—	9.0a
I_3	11.0	—	11.5	—
C	16.0	—	22.0	15.0
P_1	19.2	17.5	18.5	17.0a
P_2	20.4	—	22.0	17.5a
P_3	20.5	—	21.5	19.2
P_4	21.2	18.0	21.2	20.0
M_1	25.2	25.0	26.6	23.5a
M_2	27.3	25.5	29.6	25.2
M_3	32.6	36.2	36.3	31.5
(Transverse diameter)				
I_1	7.1	—	7.8	6.0
I_2	8.4	—	—	8.0
I_3	9.3	—	10.0	—
C	14.6	10.0	16.0	17.0
P_1	11.0	—	11.0	9.7
P_2	15.6	—	15.4	14.4
P_3	17.3	—	18.2	16.1
P_4	18.0	16.8	19.7	16.5
M_1, trigonid	18.3	19.0	22.0	17.9
M_2, trigonid	20.2	21.3	23.0	18.0a
M_3, trigonid	19.1	21.0	21.6	17.0a
M_1, talonid	18.9	18.0	21.0	17.9
M_2, talonid	19.4	18.0	22.0	16.9
M_3, talonid	15.5	14.0	17.7	12.8
Length from anterior margin of P_1 to posterior margin of P_4..	74.0	63.0a	89.0	71.5
Length from anterior margin of M_1 to posterior margin of P_4	84.0	79.4	92.0	80.0a
Length from anterior margin of canine to posterior margin of M_3....................	172.0a	—	206.0a	181.0a

REFERENCES

BROWN, R. W. 1948. Correlation of the Sentinel Butte shale in Western North Dakota. *Bull. Amer. Assoc. Pet. Geol.* **32**(7) : 1265–1274.

CAILLEUX, A. 1945. *Coryphodon*, européens et américains. *Mammalia* (Paris) **9**(2) : 33–46.

CHARDIN, P. TEILHARD DE. 1927. Les mammiferes de l'Eocene Inferieur de la Belgique. *Mem. Mus. Roy. Hist. Nat. Belgique* **36** : 1–33.

CHARDIN, P. TEILHARD DE, AND C. C. YOUNG. 1936. A Mongolian amblypod in the Red Beds of Ichang (Hupeh). *Bull. Geol. Soc. China* **15**(2) : 217–223.

CHOW, M. 1957. A new *Coryphodon* from Sintai, Shantung. *Vertebrata Palasiatica* **1**(4) : 301–304.

CHOW, M., AND CHANG-KANG HU. 1956. The occurrence of the [a] Paleogene mammal in Sinkiang. *Acta Palaeontologica Sinica* **4**(2) : 239–241.

COPE, E. D. 1872a. On Bathmodon an extinct genus of ungulates. *Proc. Amer. Philos. Soc.* **12** : 417.

——. 1872b. Second notice of extinct vertebrates from Bitter Creek, Wyoming. *Proc. Amer. Philos. Soc.* **12** : 487.

——. 1873. On the short footed *Ungulata* of the Eocene of Wyoming. *Proc. Amer. Philos. Soc.* **13** : 38–74.

——. 1874. Report upon vertebrate fossils discovered in New Mexico, with description of new species. *Geogr. Expl. and Surv. W. of 100th Merid., Appendix FF, Ann. Rep. Chief of Engineers for 1874*, 1–18.

——. 1875. Systematic catalogue of Vertebrata of the Eocene of New Mexico, collected in 1874. *Geogr. Expl. and Surv. W. of 100th Merid.*, 1–37.

——. 1877a. On the brain of *Coryphodon*. *Proc. Amer. Philos. Soc.* **16**(99) : 616–620.

——. 1877b. Report upon the extinct Vertebrata obtained in New Mexico by parties of the expedition of 1874. The Amblypoda. *Rep. U. S. Geogr. Surv. W. of 100th Merid., Paleontology part 2*, 206–209.

——. 1882. Two new genera of the Puerco Eocene. *Amer. Nat.* **16** : 417–418.

——. 1883a. The ancestor of *Coryphodon*. *Amer. Nat.* **17** : 406–407.

——. 1883b. Some new Mammalia of the Puerco formation. *Amer. Nat.* **17** : 968.

——. 1884. The Amblypoda. *Amer. Nat.* **18** : 1110–1121, 1192–1202.

——. 1888. Synopsis of the Vertebrate fauna of the Puerco Series. *Trans. Amer. Philos. Soc.* **16** : 298–361.

DOUGLASS, E. 1902a. The discovery of Torrejon mammals in Montana. *Science*, N. S., **15** : 272–273.

——. 1902b. A Cretaceous and lower Tertiary section in south-central Montana. *Proc. Amer. Philos. Soc.* **41** :207–224.

——. 1908. Vertebrate fossils from the Fort Union beds. *Ann. Carnegie Mus.* **5** : 11–26.

EARLE, C. 1892. Revision of the species of *Coryphodon*. *Bull. Amer. Mus. Nat. Hist.* **4**(1) : 149–166.

EDINGER, T. 1950. Frontal sinus evolution (particularly in the Equidae). *Bull. Mus. Comp. Zool.* **103** : 409–496.

——. 1956. Objets et résultats de la Paléoneurologie. *Ann. Palaeont.* **42** : 97–116.

FLEROW, C. C. 1952. Pantodonty (Pantodonta), sobrànnye mongol'skoï paleontologicheskoï ekspeditsieï Akademii Nauk SSSR. *Paleontologicheskiĭ Institut. Trudy* **41** : 43–50.

——. 1957. A new coryphodont from Mongolia, and on evolution and distribution of the Plantodonta. *Vertebrata Palasiatica* **1**(2) : 73–81.

FORSTER-COOPER, C. 1932. On some mammalian remains from the lower Eocene of the London Clay. *Ann. and Mag. of Nat. Hist.*, series 10, **9** : 458–467.

GAZIN, C. L. 1936. A taeniodont skull from the lower Eocene of Wyoming. *Proc. Amer. Philos. Soc.* **76**(5) : 597–612.

——. 1952. The lower Eocene Knight formation of western Wyoming and its mammalian faunas. *Smithsonian Misc. Coll.* **117**(18) : 1–82.

——. 1953. The Tillodontia. An early Tertiary order of mammals. *Smithsonian Misc. Coll.* **121**(10) : 1–110.

——. 1956. Paleocene mammalian faunas of the Bison Basin in south-central Wyoming. *Smithsonian Misc. Coll.* **131**(6) : 1–57.

GIDLEY, J. M. 1917. Notice of a new Paleocene mammal, a possible relative of the titanotheres. *Proc. U. S. Nat. Mus.* **52** : 431–435.

GRANGER, W. 1917. Notes on Paleocene and lower Eocene mammal horizons of northern New Mexico and southern Colorado. *Bull. Amer. Mus. Nat. Hist.* **37** : 821–830.

GRANGER, W., AND W. K. GREGORY. 1934. An apparently new family of Amblypod mammals from Mongolia. *Amer. Mus. Nov.* **720** : 1–8.

HÉBERT, E. 1856. Recherches sur la fauna des premiers sédiments tertiares Parisiens (mammifères pachydermes du genre *Coryphodon*). *Ann. Soc. Naturelles.* **4**(6) : 87–136.

JEPSEN, G. L. 1930a. New vertebrate fossils from the lower Eocene of the Bighorn Basin, Wyoming. *Proc. Amer. Philos. Soc.* **69** : 117–131.

——. 1930b. Stratigraphy and paleontology of the Paleocene of northeastern Park County, Wyoming. *Proc. Amer. Philos. Soc.* **69** : 463–528.

——. 1940. Paleocene faunas of the Polecat Bench formation, Park County, Wyoming. *Proc. Amer. Philos. Soc.* **83**(2) : 217–340.

KAMPEN, P. N. VAN. 1905. Die tympanalgegend des saugetierschadels. *Morpholog. Jahrbuch* **34** : 321–722.

KLAAUW, C. J. VAN DER. 1931. The auditory bulla in some fossil mammals. *Bull. Amer. Mus. Nat. Hist.* **62** : 1–352.

LLOYD, E. R., AND C. J. HARES. 1915. The Cannonball marine member of the Lance formation of North and South Dakota and its bearing on the Lance-Laramie problem. *Jour. Geol.* **23** : 523–547.

MARSH, O. C. 1876. Characters of the genus *Coryphodon*. *Amer. Jour. Sci. and Arts* **2** : 425–428.

——. 1877. Principal characters of the Coryphodontidae. *Amer. Jour. Sci. and Arts* **14** : 81–88.

——. 1893. Restoration of *Coryphodon*. *Amer. Jour. Sci. and Arts* **46** : 321–326.

MATTHEW, W. D. 1897. A revision of the Puerco fauna. *Bull. Amer. Mus. Nat. Hist.* **9** : 259–323.

——. 1937. Paleocene faunas of the San Juan Basin, New Mexico. *Trans. Amer. Philos. Soc.* **30** : 1–510.

MATTHEW, W. D., AND W. GRANGER. Fauna and correlation of the Gashato formation of Mongolia. *Amer. Mus. Nov.* **189** : 1–12.

OSBORN, H. F. 1898a. A complete skeleton of *Coryphodon radians*. Notes upon the locomotion of this animal. *Bull. Amer. Mus. Nat. Hist.* **10** : 81–91.

——. 1898b. Evolution of the Amblypoda. Part I. Taligrada and Pantodonta. *Bull. Amer. Mus. Nat. Hist.* **10** : 169–218.

——. 1909. Cenozoic mammal horizons of western North America with faunal lists of the Tertiary mammalia of the West. *U. S. Geol. Surv. Bull.* **361** : 1–438.

——. 1924. *Eudinoceras*. Upper Eocene Amblypod of Mongolia. *Amer. Mus. Nov.* **145** : 1–5.

——. 1929. The titanotheres of ancient Wyoming, Dakota, and Nebraska. *U. S. Geol. Surv. Monograph* **55** : 1–953.

OSBORN, H. F., AND J. L. WORTMAN. 1892. Fossil mammals of the Wasatch and Wind River beds. Collection of 1891. *Bull. Amer. Mus. Nat. Hist.* **4** : 81–147.

OSBORN, H. F., AND C. EARLE. 1895. Fossil mammals of the Puerco series. *Bull. Amer. Mus. Nat. Hist.* **7** : 1–70.

OSBORN, H. F., AND W. GRANGER. 1931. Coryphodonts of Mongolia, *Eudinoceras mongoliensis* Osborn, *E. kholobolchiensis* sp. nov. *Amer. Mus. Nov.* **459**: 1–13.

OSBORN, H. F., AND W. GRANGER. 1932. Coryphodonts and uintatheres from the Mongolian Expedition of 1930. *Amer. Mus. Nov.* **552**: 1–16.

OWEN, R. 1845. Odontography, or a treatise on the comparative anatomy of the teeth, etc. **1**: i–lxxiv, 1–655. London.

——. 1846. A history of British fossil mammals and birds. 299–308. London.

PATTERSON, B. 1933. A new species of the amblypod *Titanoides* from western Colorado. *Amer. Jour. Sci.* **25**: 415–425.

——. 1934. A contribution to the osteology of *Titanoides* and the relationships of the Amblypoda. *Proc. Amer. Philos. Soc.* **73**: 71–101.

——. 1935. Second contribution to the osteology and affinities of the Paleocene amblypod *Titanoides*. *Proc. Amer. Philos. Soc.* **75**: 143–162.

——. 1936. Mounted skeleton of *Titanoides* with notes on the associated fauna. *Proc. Geol. Soc. Amer. 1935*: 397–398.

——. 1937. A new genus Barylambda, for *Titanoides faberi*, Paleocene amblypod. *Field Mus. Nat. Hist., Geol. Ser.* **6**(16): 229–231.

——. 1939a. New Pantodonta and Dinocerata from the upper Paleocene of western Colorado. *Field Mus. Nat. Hist., Geol. Ser.* **6**(24): 351–384.

——. 1939b. A skeleton of *Coryphodon*. *Proc. New England Zool. Club* **17**: 97–110.

——. 1949. Rates of evolution in taeniodonts. Genetics, paleontology and evolution. Princeton, Princeton University Press.

PATTERSON, B., AND E. L. SIMONS. 1958. A new barylambdid pantodont from the late Paleocene. *Brev. Mus. Comp. Zool.* **93**: 1–8.

PAULA COUTO, C. DE. 1952. Fossil mammals from the beginning of the Cenozoic in Brazil. Condylarthra, Litopterna, Xenungulata, and Astrapotheria. *Bull. Amer. Mus. Nat. Hist.* **99**(6): 359–394.

RUSSELL, L. S. 1948. A middle Paleocene mammal tooth from the foothills of Alberta. *Amer. Jour. Sci.* **246**: 152–156.

SCOTT, W. B. 1937a. The Astrapotheria. *Proc. Amer. Philos. Soc.* **77**(3): 309–393.

——. 1937b. A history of land mammals in the Western Hemisphere. Revised edition. New York, Macmillan.

SIMPSON, G. G. 1929. A new Paleocene Uintathere and molar evolution in the Amblypoda. *Amer. Mus. Nov.* **387**: 1–9.

——. 1935a. The Tiffany fauna, upper Paleocene. I. Multituberculata, Marsupalia, Insectivora, and ?Chiroptera. *Amer. Mus. Nov.* **795**: 1–19.

——. 1935b. The Tiffany fauna, upper Paleocene. II. Structure and relationships of *Plesiadapis*. *Amer. Mus. Nov.* **816**: 1–30.

——. 1935c. The Tiffany fauna, upper Paleocene. III. Primates, Carnivora, Condylarthra, and Amblypoda. *Amer. Mus. Nov.* **817**: 1–28.

——. 1935d. New Paleocene mammals from the Fort Union of Montana. *Proc. U. S. Nat. Mus.* **83**(2981): 221–244.

——. 1937a. The Fort Union of the Crazy Mountain field, Montana and its mammalian faunas. *U. S. Nat. Mus. Bull.* **169**: 1–287.

——. 1937b. Additions to the upper Paleocene fauna of the Crazy Mountain field. *Amer. Mus. Nov.* **940**: 1–15.

——. 1937c. Notes on the Clark Fork, upper Paleocene, fauna. *Amer. Mus. Nov.* **954**: 1–24.

——. 1941. Large Pleistocene felines of North America. *Amer. Mus. Nov.* **1136**: 1–27.

——. 1945. The principles of classification and a classification of mammals. *Bull. Amer. Mus. Nat. Hist.* **85**: 1–350.

——. 1953. The major features of evolution. New York, Columbia University Press.

SINCLAIR, W. J., AND W. GRANGER. 1914. Paleocene deposits of the San Juan Basin, New Mexico. *Bull. Amer. Mus. Nat. Hist.* **33**: 297–316.

VAN HOUTEN, F. B. 1945. Review of latest Paleocene and early Eocene mammalian faunas. *Jour. Paleo.* **19**(5): 421–461.

WILSON, R. W. 1956. A new multituberculate from the Paleocene Torrejon fauna of New Mexico. *Trans. Kans. Acad. Sci.* **59**(1): 76–84.

WOOD, H. E. 1923. The problem of the *Uintatherium* molars. *Bull. Amer. Mus. Nat. Hist.* **48**: 599–604.

WOOD, H. E., *et al.* 1941. Nomenclature and correlation of the North American continental tertiary. *Bull. Geol. Soc. Amer.* **52**: 1–48.

WORTMAN, J. L. 1897. The Ganodonta and their relationship to the Edentata. *Bull. Amer. Mus. Nat. Hist.* **9**: 59–110.

Fɪɢ. 10.   Pantolambdid and titanoideid upper dentitions.

A. *Pantolambda bathmodon* Cope. Upper left dentition of USNM 15408, reversed.   Outline of canine and incisors from AMNH Nos. 16553 and 15554.   × 3/2.

B. *Pantolambda cavirictus* Cope.   Upper left dentition of USNM 21327, reversed.   × 1.

C. *Caenolambda jepseni* sp. nov. Upper right dentition of PU 14863.   × 1.

D. *Titanoides gidleyi* Jepsen.   Upper left dentition, reversed.   × 1.

E. *Titanoides simpsoni* sp. nov. Upper right dentition of AMNH 35720.   × 3/2.

F. *Titanoides primaevus* Gidley. Upper right dentition of PU 16490 (type), incisors omitted.   × 1.

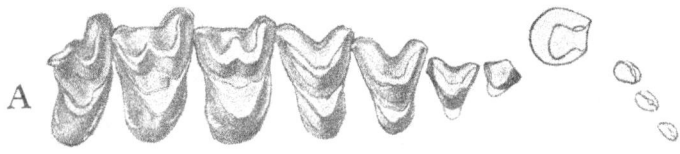

A

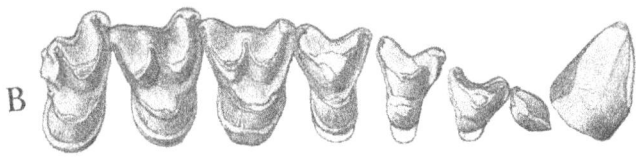

B

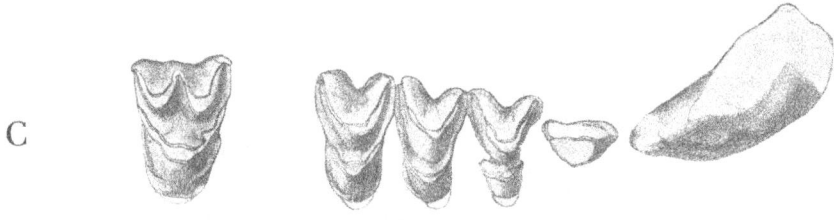

C

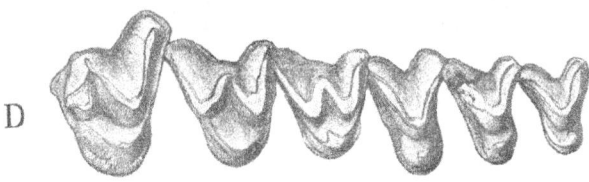

D

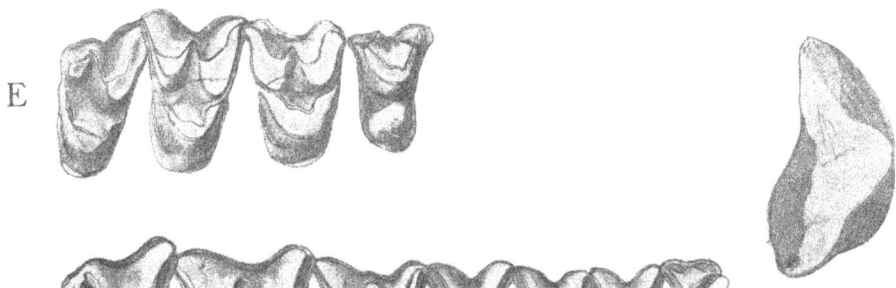

E

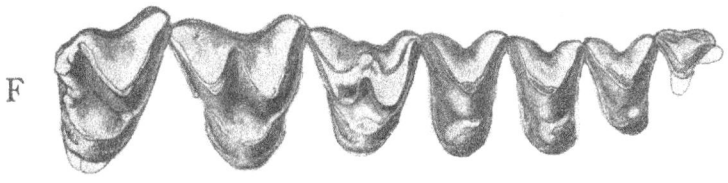

F

FIG. 11. Barylambdid upper dentitions. × 1.

A. *Barylambda faberi* Patterson. Upper right dentition of CNHM P-15075; incisors, canine, and P³ restored from other specimens principally CNHM P-14944.

B. *Leptolambda schmidti*, Patterson and Simons. Upper right dentition of PU 14996.

C. *Haplolambda quinni* Patterson. Upper right dentition; canine, P¹ and P² reversed from left side, CNHM P-15542.

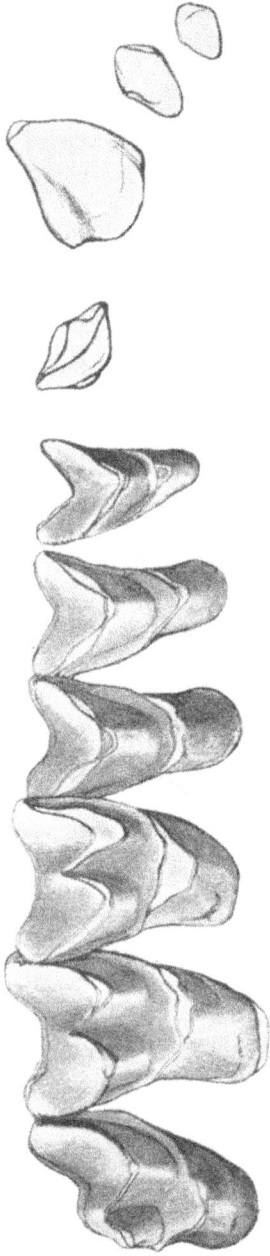

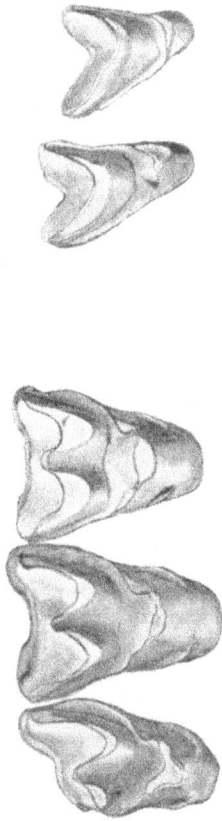

A   B   C

FIG. 12.   The dentition of *Ignatiolambda barnesi*, gen. et
sp. nov., AMNH 55400.   × 1.

A. Upper right dentition as found.

B. Upper left dentition as found.

C. Restoration of upper right dentition, based on both sides.

D. Lower right dentition as found.

E. Lower right dentition, restored from both sides.

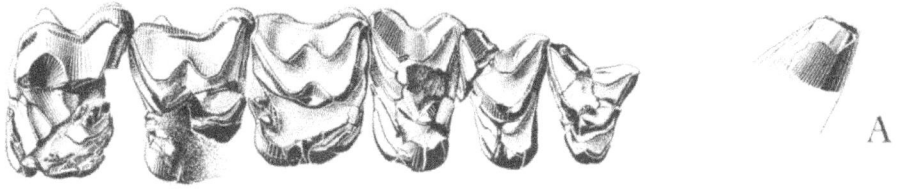

A

B

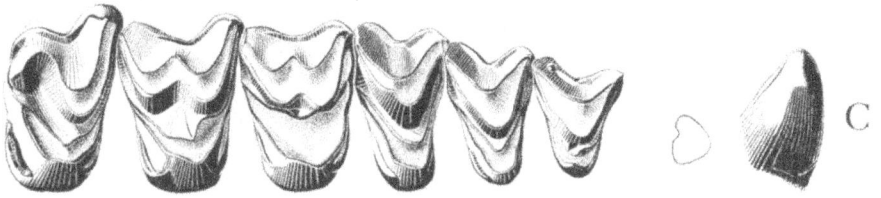

C

D

E

Fig. 13. Comparison of barylambdid and
pantolambdid mandibles. × 1/2.

A. *Leptolambda schmidti* Patterson and Simons. PU 14990,
lateral view of mandible and lower dentition.

B. *Caenolambda jepseni* sp. nov., PU 14863, lateral view of
mandible and lower dentition; P¹ restored from left side.

C. *Caenolambda jepseni* sp. nov. PU 14863, crown view of
mandible and lower dentition; right P¹ restored from left
side; left P² restored from right side.

D. *Leptolambda schmidti* Patterson and Simons. PU 14990,
crown view of mandible and lower dentition.

A

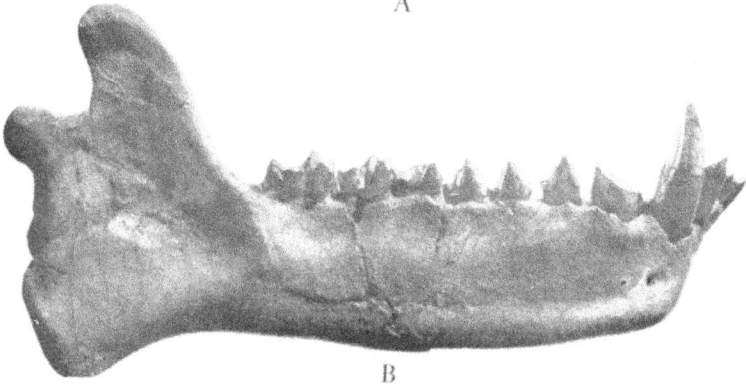

B

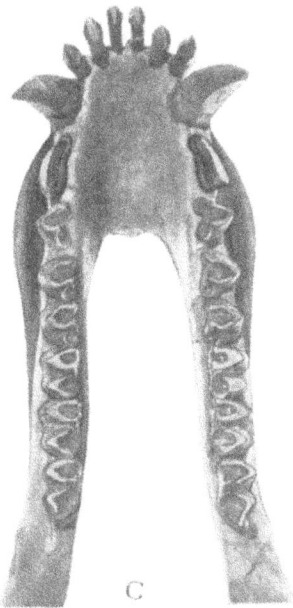

C

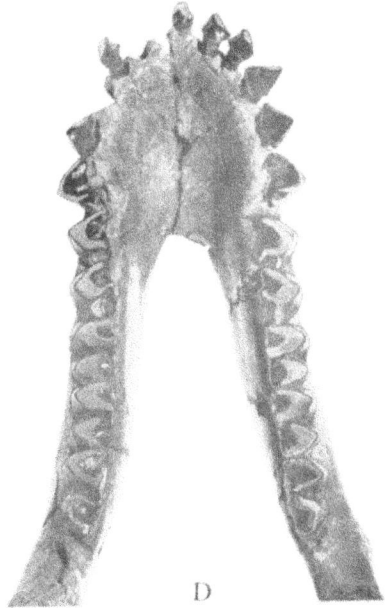

D

FIG. 14. Ventral view of a titanoideid skull. × 1/2. *Titanoides primaevus* Gidley. Dentition and premaxilla restored from PU 16490 (type) of upper right side, and reversed on left. Details of palate and glenoid fossa of squamosal added from CNHM P-15520. Skull, based on USNM No. 20029 posterior to dotted line at *a* with slight correction for distortions.

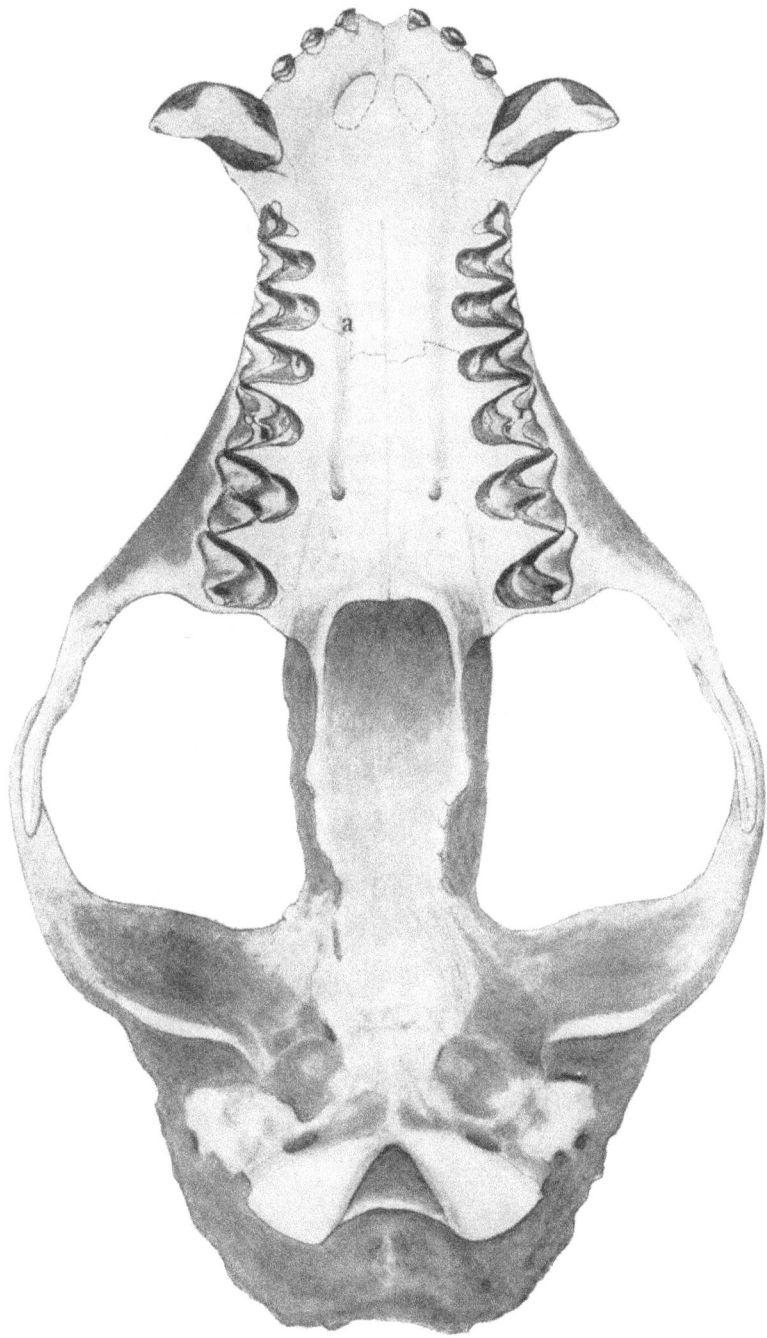

Fig. 15. The petrosal of *Titanoides*.

A. *Titanoides primaevus* Gidley.   Photograph of ventral aspect
   of right *os petrous* of PU 16490 (type).   Scale about twice
   natural size.   Tympanophyal restored from CNHM P-15520.

f.s.	stylomastoid foramen.
d.f.n.	dehiscence of facial nerve.
fs.o.	fenestra ovale.
p.c.	promentorium cocleae.
fs.r.	fenestra rotunda.
g.st.	groove for stapedius.
d.aur.n.	dehiscence of auricular nerve.
s.c.t.	support for corda tympani.
th.	tympanohyal.

B. *Titanoides primaevus* Gidley.   Generalized diagram of basicranium of *Titanoides* based on USNM No. 20029, showing
   position of *os petrous* figured above.   Scale about one half
   natural size.

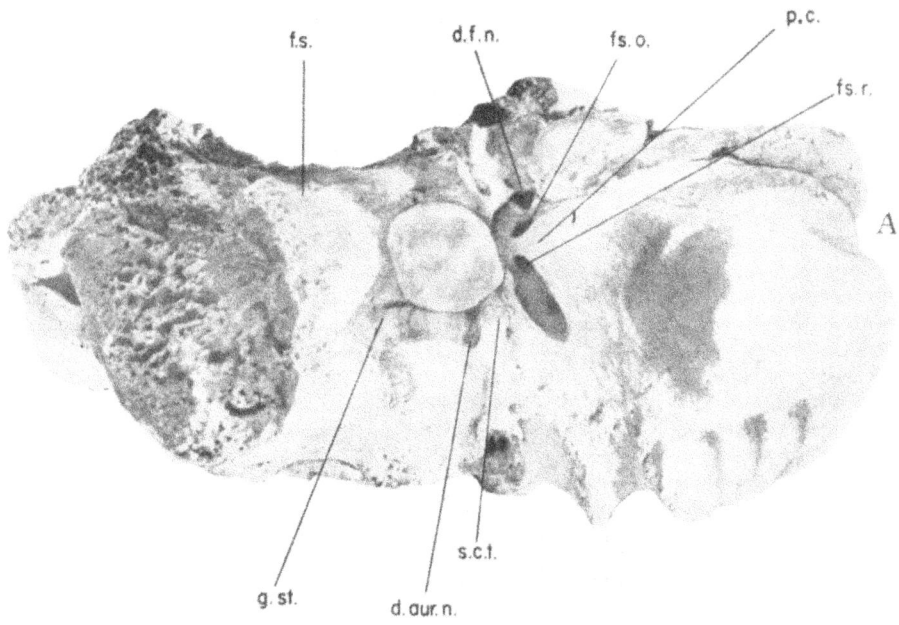

f.s. d.f.n. fs.o. p.c.

fs.r.

A

g.st. d.aur.n. s.c.t.

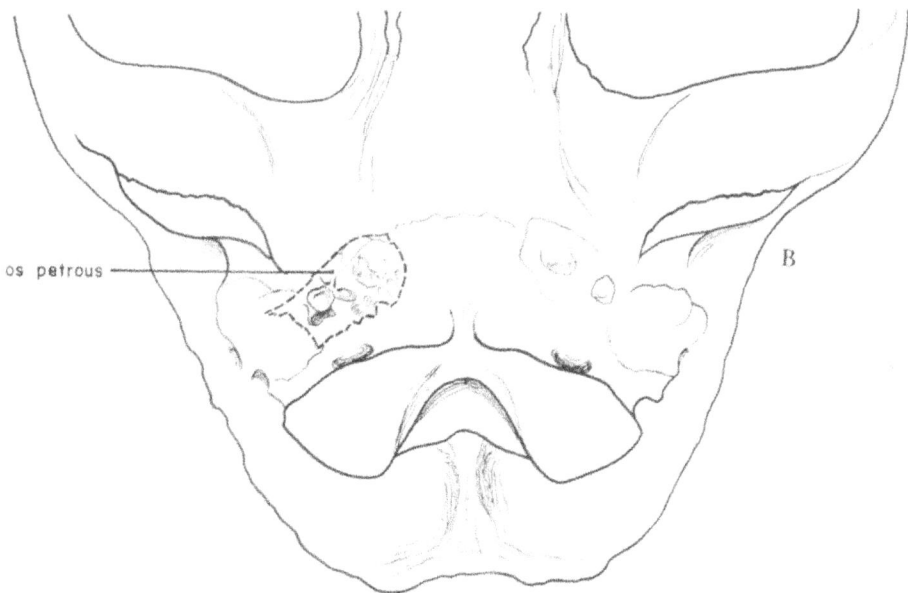

os petrous

B

93

FIG. 16. Pelves of Paleocene pantodonts.

A. *Pantolambda bathmodon* Cope.   Pelvis and sacrum of AMNH 16663.   × 1/3.

B. ? *Pantolambda cavirictus* Cope.   Pelvis of AMNH 2455.   × 1/4.

C. *Coryphodon* sp.   Pelvis of PU 14685 of Clarkforkian age or earlier.   × 1/5.

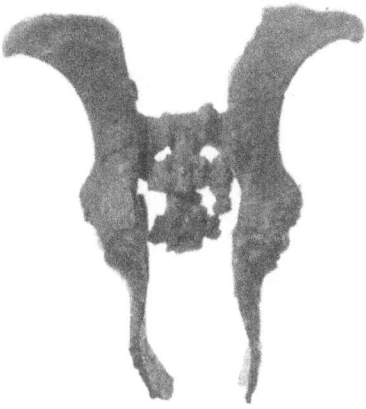

A

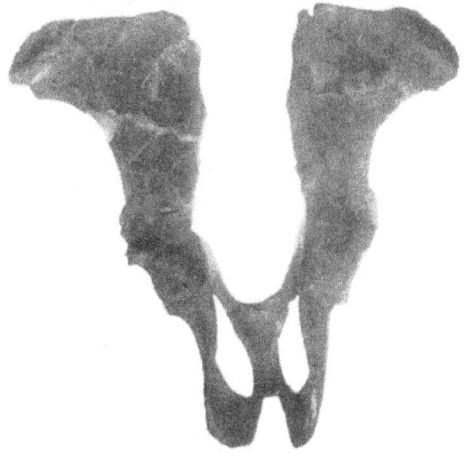

B

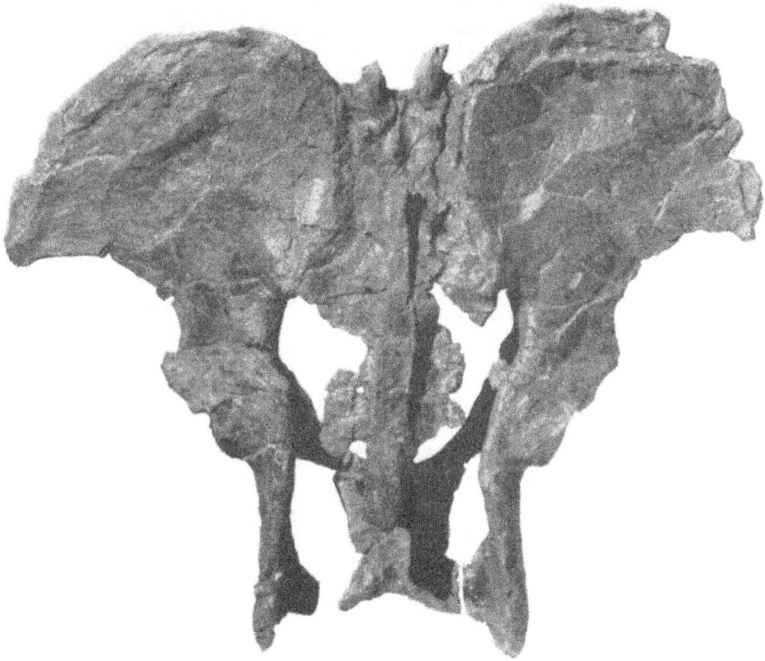

C

Fig. 17. Photographs showing presumed paratypes of *Panto-lambdodon inermis* belong to one individual. × 7.

A. *Pantolambdodon inermis* Granger and Gregory.   Unretouched photograph of the anterior face of the paratype last lower molar, AMNH No. 21748.   Pointer indicates worn surface which exactly matches that on posterior face of lower molar of AMNH No. 22100, shown below.

B. *Pantolambdodon inermis* Granger and Gregory.   Unretouched photograph of posterior face of the last lower molar preserved in lower jaw fragment of the paratype specimen AMNH No. 22100.   Pointer indicates worn surface, which shows that this tooth preceded molar illustrated in A.

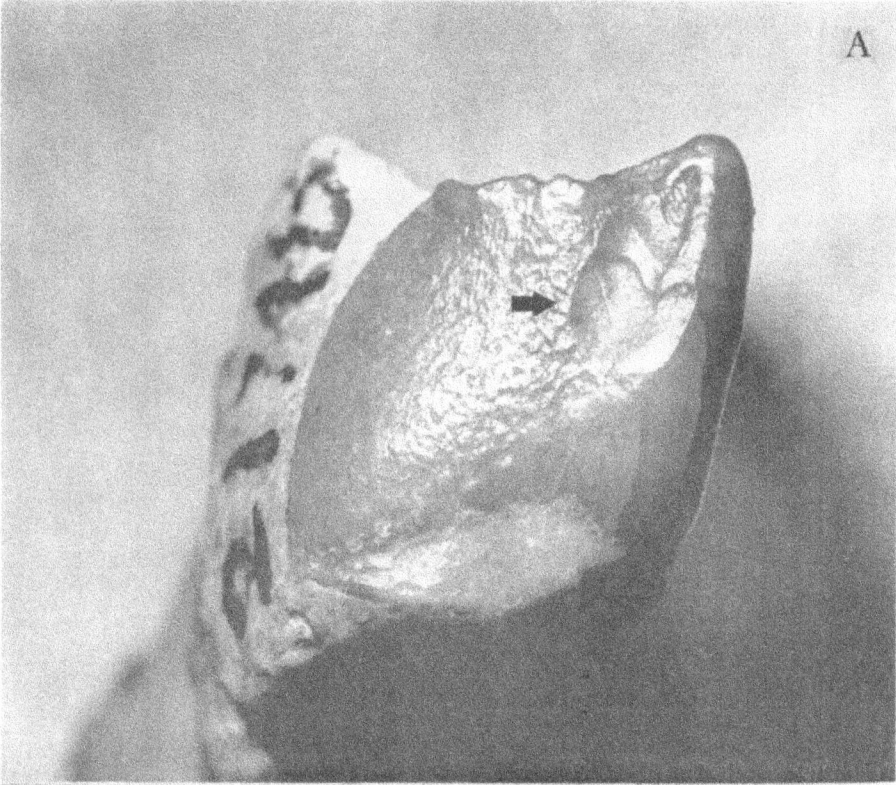

FIG. 18. Two new pantodont species. $\times$ 7/10.

A. *Coryphodon proterus* sp. nov. PU 13400, crown view of partial lower dentition.

B. *Titanoides majus* sp. nov. PU 16447, crown view of P_{2-4} and anterior horizontal ramus of mandible.

C. *Titanoides majus* sp. nov. PU 16447, internal aspect of mandibular ramus showing region of symphysis and P_{2-4}. (Stippling indicates missing or damaged areas.)

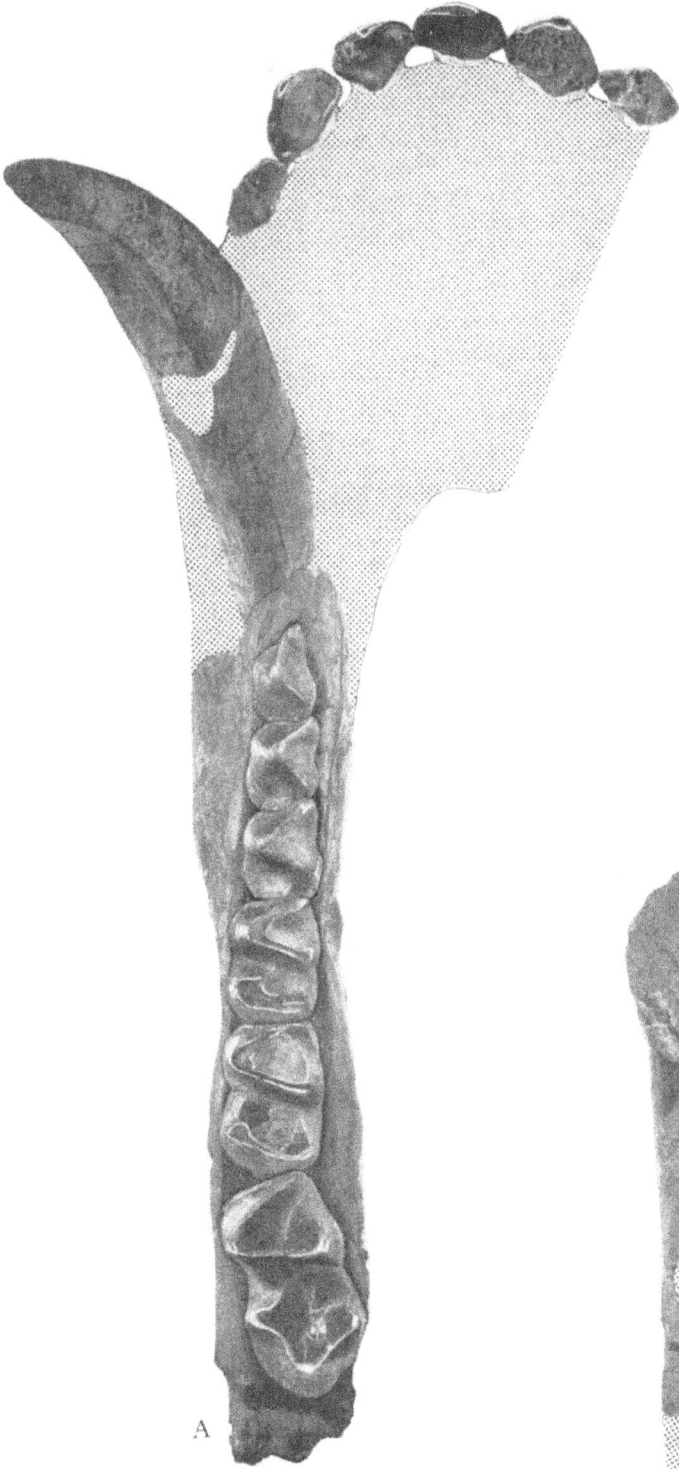

www.ingramcontent.com/pod-product-compliance
Lightning Source LLC
Chambersburg PA
CBHW081336190326
41458CB00018B/6025